Microgenetic Theory and Process Thought

Jason W. Brown

imprint-academic.com

Copyright © Jason W. Brown, 2015

The moral rights of the authors have been asserted.
No part of this publication may be reproduced in any form
without permission, except for the quotation of brief passages
in criticism and discussion.

Published in the UK by
Imprint Academic, PO Box 200, Exeter EX5 5YX, UK

Distributed in the USA by
Ingram Book Company,
One Ingram Blvd., La Vergne, TN 37086, USA

ISBN 9781845407698

A CIP catalogue record for this book is available from the
British Library and US Library of Congress

Contents

Preface	v
Chapter 1. The Mind/Brain State: Nature	1
Chapter 2. The Mind/Brain State: Origins	16
Chapter 3. Experience	30
Chapter 4. What is Consciousness?	47
Chapter 5. Feeling	65
Chapter 6. Thinking	87
Chapter 7. Novelty and Causation	114
Chapter 8. Certainty and Conviction	134
Chapter 9. Psychology of Free Will	153
Theoretical Note 1. Probability and Potential	171
Theoretical Note 2. On Truth	178
Thoeretical Note 3. On the Direction of Time	187
Theoretical Note 4. Origins of Religious Belief	195
References	202
Index	211

*Now that my ladder's gone,
I must lie down where all the ladders start
In the foul rag and bone shop of the heart.*
—W.B. Yeats, *The Circus Animal's Desertion*

Preface

I
Theoretical Background

The chapters in this collection attempt to establish, on a psychological basis, some foundational principles of a philosophy of mind that are grounded in process (microgenetic) theory and evolutionary principles. The approach diverges widely from conventional methods in that the contextual and historical (diachronic) aspects of mental contents and the mind/brain states in which they are ingredient are the primary focus of investigation, instead of treating them as logical solids and debating their assumed properties. In my view, in spite of the massive literature that has accumulated over forty years of research in cognitive science, psychology made a wrong turn implementing a methodology to mimic physical science and reinforce linguistic and analytic philosophy, namely, the study of mental components in isolation, conceived as nodes in a network or circuit-board, and aided by imaging techniques that purport to justify the psychological method by weak, imaginary or fictitious correlations. In experimental cognition, the locus and connectivity of mental contents replace the transition and temporality in which they are embedded. The result is that the dynamic and subjectivity that flow from the underpinnings of conscious experience are shredded, and the psyche is conceived as a compilation of interconnected sub-routines.

For microgenesis, the micro-temporal process leading to a conscious endpoint is, together with the final content, part of an epochal state. The outcome of a state, such as an object or word, is not a resultant of the preceding series but incorporates earlier segments of that series — value, meaning, belief — as part of what it is. An object includes its formative phases. The subjective has inner and outer segments. The world is the surface of the mental state. Final actualities specify pre-object phases which detach and articulate mind-external. Some remarks of Stout (1902) are germane to the current impasse. "Psychology investigates the history of individual consciousness, and this coincides with the history of the process through which the world comes to be presented in consciousness... The aim of psychology is purely retrospective... to go back upon the traces of experience, and ascertain how [an] existing standpoint has arisen... When, on the other hand, the nature of knowledge is considered apart from its genesis, and in abstraction from its time-vicissitudes, it becomes the subject-matter, not of psychology, but of metaphysics." The aim of this book, as James might have put it, is for a tidal wave of process thought to surge upon the dry shoals of metaphysical speculation.

One salutary effect of process-thinking is a return to continuities between diverse aspects of cognition that, once fractured by analysis, become engraved in the mind and solidified by empirical or quantitative study, all in the hope of recombination once dismemberment is complete. The vanity of the cognitivist paradigm is exposed in the attempt to unify a multitude of isolates by the *ad hoc* postulate of a "binding mechanism" to repair the failed heuristics of demarcation, interaction and external relations. In scientific thought, wholes are sums of parts. In process thought, wholes are potentials for virtual partitions or categories for specification. Parts are not *in situ* in wholes but are novel derivations that can serve as sub-categories for ensuing partitions. A qualitative account of wholes and parts from a "genetic" perspective (Brown, 2005a) provides an alternative to the causal theory of the mind/brain in that it reveals a correspondence between brain and mental process over correlated phases, such that the mental state can be conceived as isomorphic with the brain state.

A causal account of mental process, including object-perception, obligates physical properties similar if not identical to those assumed for brain. While causation in "mental process" enjoys a healthy uncertainty, the same reservations should apply to brain process. The term "mental process" or "mental activity", liberally used in many writings, including to some extent the present one, tends to denote causation. In contrast to brain study, where stimulation of neurons induces an effect on other neurons, or on behavior, e.g. limb movement, or on psyche, e.g. an epileptic aura, the immediate experience of mental process is not of antecedence and subsequence. A thought may be followed by another thought, but thoughts are not necessarily coupled and decision does not invariably lead to action. Except for the feeling of activity and passivity, or agency and receptiveness, there is at most the feeling of a dynamic that underlies final contents, a feeling that is accentuated when the direction of activity is impeded, as in tension, hesitation or anxiety. If we could eliminate acts, objects or mental contents in a momentary cognition, mental activity would likely be felt as pure feeling without origin or subjective aim. The lack of direction or intentionality would suspend the feeling of before and after and result in a felt stasis of energy.

An account of the mind/brain invites an interpretation of change. In contrast to the common sense version of "apperception", change in the object world is not perceived directly but is a consequence of changing conditions that are mirrored in, or realized through, overlapping reinstatements. As in the phi phenomenon, one mind/brain state is substituted for another in rapid succession, such that illusory change is created by the imposition of ensuing states on preceding ones. The overlap gives the continuity; the replacement gives the appearance of change. In this respect, the thought of Bradley is close to the claims of this book, in which the mental state and other objects are described in terms of imminent causation, i.e. self-caused change or "causal persistence", as in the moment-to-moment replacement of objects and organisms. The similarities with the arising and perishing of point-instants in Buddhist philosophy should not escape the careful reader.

The principle axis of speculation is the theory of the mind/brain state, which is an outcome of clinical studies described in prior works, most recently Brown (2010a). To my knowledge, the microgenetic account is the only coherent and detailed theory of the mental state. It is also far removed from the tendency of philosophers to identify it with content—quale, mental statement—at a given moment. Without a theory of the mind/brain state, on which the relation of mind to brain is conditioned, philosophy can do little more than thrash about in search of adequate definitions or possible solutions but will fail to progress to explanatory closure by argument alone, especially given the deference to computational models and the resistance to subjectivist theory. The problem cannot be properly understood apart from the trajectory of the mental state and how the phase-transition of the state maps to brain process. The difficulty is compounded by uncertainty as to mental contents and properties, the presumption of causal brain process, possibly causal mental process, and the epistemic status of perceptual objects. The reader immune to internalist theory, or refractory to explanation in terms of antecedent process, or to recurrence and epochal states, will be forgiven if he closes the book at this point.

This work was conceived as a whole, though some chapters have appeared as journal articles. This has the effect, perhaps helpful to some, perplexing to others, of a repetition of theory and arguments. The justification for each chapter standing alone with slight revision of the original is that it can be more easily related to the general theory instead of scavenging the chapters to accommodate a linear exposition. It was felt there would be value for each chapter to be read independent of the others in spite of the overlap since every topic in the book owes to a common source. As Yeats wrote, "though leaves are many, the root is one." The sole disclaimer is that a reader unwilling to look up the cited sources must accept on faith much of the groundwork for the theory, since the data that gave birth to it consist largely of patterns of error (symptom) formation in neuropsychological cases with focal brain lesions.

II
Metaphysical Principles

The nature of perception and the perception of nature, the subjective and objective, appearance and reality, ancient problems in philosophical discourse, are central topics in this work. Here, however, conclusions are not reached by logic and argument alone, or by introspection and intuition but, as mentioned, by the evidence of clinical studies, which have all been described in past writings. The prior work forms a coherent system of thought that offers a novel path into these perennial issues. The pathological data are critical in revealing the microstructure of subjectivity, for which argumentation has little consequence, and to which introspection has limited access. The data confirm that spatiotemporal events in the world are, in fact, appearances in the mind, and that perceptions are not assembled or constructed into mental objects and projected outside, but are exact representations or conceptual models of a hidden reality. Object features and object relations are not themselves the source of their perception but are analytic outcomes of a process of image-creation. Objects arise as endogenous images constrained by physical sensibility to mirror or model the unobservable. Object-relations and the temporal order of events in the world are endpoints in the actualization of the mental state. Brain and object are presumed to refer to physical entities that are indirectly known through the study of brain and object-perception.

An object-appearance does not occur in isolation but is embedded in, and presupposes, a world of object-relations. To perceive space or to apprehend time is to generalize spatial and temporal particulars within a totality of appearances. The world of perceived objects and object-relations is a mental image in its entirety that can be experienced as a whole or as parts of a whole depending on the locus in a progression from psychic categories to world-close concepts, as well as on the attention or value given to the parts. There are no differences in the perceived change in an object and the change in an object perceived apart from the attention (value) given to a locus in the perceptual field. An object is a feature in its surround, abstracted from change and isolated in an event-series through a conceptual analysis over the actualization process. The spatial field of one individual relates to that of another in the same manner as one mind relates to another, though the space of an observer conforms to the space of others by virtue of a common field of sense-data, while the subjectivity of the temporal field, because it is less impacted by immediate sense-data, has less overlap with other minds and greater individuality. Each person creates a model of the world that conforms (adapts) to sensibility, a model that must be relatively precise to accommodate the laws of mathematics, indeed, for survival in an imaginary world. The model of other minds is less objective. We can safely navigate the space of external objects but only in a general way can we sort out the thoughts and dispositions of other minds.

Appearances derive from the core self in a passage from categorical primitives to refined concepts. The object is specified out of the self, which anchors the state in a relation of an observer to a succession of images. The self is an "on-looker" from a locus early in the mental state to its terminus over an intentional arc essential for consciousness. For many thinkers, from the Anatman theory of Buddhism to David Hume and William James (discussion in Bricklin, 2014), there have been doubts as to the validity of the self-concept, or whether it makes sense to speak of a self, even if "self-ascription" is a necessary accompaniment of having a conscious experience. The feeling this is "my" experience, or of the "I" in action, is a vital part of having a conscious experience at all. In fact, the self as source of concepts, feelings, acts, objects, i.e. the passage of all experience through the self, entails self-ascription even for unconscious events, such as subliminal perception or incidental learning. Feelings and concepts are not attached to objects but embody them. The self — not necessarily the self of waking consciousness — is engaged even if, as in dream, intentionality is incomplete and an object is not realized.

The self is not an object of consciousness like other objects; it is ingredient in the state and, like a thought, can be a topic or focus of attention. A self as a conscious (pre)object is not comparable to a self that is conscious of objects or ideas, for it supposes a reflexive attitude in a process that is unidirectional. The self is the origin not the aim of a state of consciousness. The core self through which the mental state develops is not the self of introspection. The former is unconscious and bound up with attributes of character; the latter is at the threshold of consciousness, bound up with agency, desire and decision. The core and conscious self, as categories and sources of the mental state, partition to inner and outer contents over phases of affective and memorial experience. For this reason, there is implicit value and recognition for all perceptual objects, even prior to conscious perception, for example, as a preparation for action. Incidentally, since misrecognition entails recognition (of unfamiliarity), the same traversal occurs with a mismatch to memory.

The presence of an object-appearance is essential for waking consciousness. The distinction of external objects and their perception amounts to a distinction of the subjective process of object-formation and the endpoint of this process in an object. That is, the relation of perceiving to what is perceived is a relation of proximal to distal in a continuous process. The relevant distinction is between the perceptual object and the sensibility and/or physical entity it represents. The need for an object for conscious experience is a need for the process to complete itself. This entails that sense-data carry the image outward, especially at the final phase of image-development, though sensibility figures at all phases, not just by impingements on the sensory surface but within the brain state. The objectivity of the world, its persistence and stability, are inferences from recurrent appearances that are specified by empirical data.

The object is the culmination—actually, the accumulation—of a sequence of potential or virtual entities, from the vague, the uncertain and the intangible to the real, the palpable and the definite. In the course of this series—a subjective continuum—we can speak of two series at each phase in process, the subjective phenomenon—overt or concealed—and its physical (or non-physical) correlates. Ordinarily, this reduces to a psychic manifestation of brain process, but there is also, historically, a distinction of an empirical or conscious self, and a self-in-itself, a Kantian self *an sich*. The self would then point to a self out of time, free of its material basis and causal history. Though a topic of derision by some philosophers, notably William James, if we can conceive of a self *an sich*, we should also conceive of a thought-in-itself, i.e. a reality that corresponds to a concept or idea. This distinction is most pronounced at the terminus of the mind/brain state in a perceptual object. The transition from self to thought, and from thought to object, can be conceived as a processing sequence over levels in brain activity, but it might also correspond to a reality uncoupled from brain process. A self or thought out of time is like an eternal idea, not dependent on a physical entity.

If the self is appearance (illusion), to what reality—if any—does it conform? Many would say none, interpreting the self as an apparition laid down in the brain state. Some have claimed the self is the product of belief, but this begs the question of the grounds of a belief. Similarly, the claim that the self is a product of Will attempts to absorb the problem in a topic no less in need of clarification. I suspect the self is a combination (category) that encloses core beliefs, values, dispositions, the will or instinct to survive, the inherited repertoire and early (unconscious) experiential memories. These are aspects of a relatively stable, i.e. recurrent, category. In referring to the self, we say I believe, value, feel, desire, when the self is a composite of the activities attributed to it. The concept of the self as category, and the will as process, resolves the origins of belief and value, or concept and feeling, as dispositional members of a single category.

A self is comparable to an idea. A self-concept and an object-concept differ primarily in the approximation to objectivity. An idea corresponds to a pattern of brain activity as well as to events in the world. What events in the world might correspond to the self, as a category, an illusion or a collection of attributes? An idea, like the self, is grounded in the brain, but an idea can also refer to something in the world. An idea is closer to the world than the self, while the object is still closer. Transitional phenomena such as eidetic images provide support for a continuum. The self has the brain as correlate. So too for thought, but thoughts also correspond to external states of affairs, while an object is apprehended as external and independent. There is progressive reference to a world outside mind and brain in the passage from self to thought to object. This invites the speculation that every phase has, to a greater or lesser extent, an external non-neural referent, e.g. self/soul, thought/state of affairs, object/real entity.

The simplest route to objectivity is to assume the perceived object is real and substantial. However, if the objective is the outer limit of subjectivity that externalizes in the adaptation to sensory negation, the object will be the conclusion of a process of object-formation. Phases in this process can precipitate as mental contents — images, propositions — that are linked to the object as segments in perceptual realization. Even on a mechanical interpretation, an object constructed of features and projected into the world is still an event in the brain. An object cannot find its way outside the observer's head.

The human mind has the capacity for a judgment as to the realness of an object, as well as to its interpretation and meaning. In most instances, the quality of realness is immediate, a feeling not a judgment. When we are uncertain, judgment compares the context and quality of realness to that of objective reality. Hallucination may seem real but it is not object-like in its realness. Indeed, as soon as the realness of the world is questioned, one's grip on reality is in danger. Realness is not reality, but the implicit feeling a phenomenon is real is essential to such a judgment. Many phenomena that are unreal, such as rainbows or holograms, are judged as unreal because they are discordant with other modalities or with the wider field of reality. Dream images are felt as real because of the coherence of modalities and the absence of critical judgment during the dream. A judgment as to whether a state of affairs is true or false concerns linguistic correlates (verbal images) of the forming object. The judgment is implicit in the object-appearance but not essential to it. When introspection seeks the meaning or interpretation of an object, earlier phases in the original perception propagate as thought. The dominant focus recedes from the object to its psychic precursors, which refer to external states of affairs and of which the object is a part. In receding from an act or object, images reappear as predecessors.

For the higher animals, objects are also appearances, while in primitive organisms the boundaries of the organic and physical are indistinct. A judgment entails a statement that is true or false. In contrast, an appearance is real or unreal, exciting, dangerous, beautiful, but not true or false. The matter of truth enters with a proposition concerning the object, not the object itself. A statement that an object is beautiful supplements the perception of its beauty. The degree of realization across perceptual modalities is also a factor in a judgment of the realness of experience. Realness is a subjective feeling; reality is an objective inference. Consistent with this account, Merleau-Ponty (1968ed, p. 16) argued against the cleavage of subjective and objective, maintaining that "the contact between the observer and the observed (should) enter into the definition of the 'real'."

It is striking that, with the exception of a few philosophers, alterations in veridical perception in dream and pathological objects, which demonstrate a continuum from subject to object, have been largely neglected. One reason for this is that the common distinction of perception from dream or hallucination is taken to be what is true and what is false, when as noted this distinction is one of the real and the unreal. The reality of an object-appearance is more secure than that of a mental image. Ordinarily, observers do not argue over the presence of a table in the room, but they do argue over statements about the table. Thought is verbal imagery in which the true and the veridical become actual when the image externalizes in speech or writing and undergoes adaptation by way of sensory revision in discussion or demonstration. The truth of a thought (image) must be defended while the reality of an external image (object) is usually not in dispute.

The presence of systems in the brain for color, movement, depth and so on supports the contention that organisms evolve in conjunction with their environment. The organism changes, and changes with, the environment. Strawson (1966) wrote in a related context that "the representation of such capacities presupposes the capacity to acquire them." The mapping of mind to world is an adaptation to the three-dimensional space of medium-sized objects in which we and most other mammals live. It is not surprising that systems in brain evolved in response to sense-data to elaborate a subjective model. It is not clear to what extent such systems are innate or acquired, though the relation of brain development to sense-data is essential for normal maturation. The distinction of appearance and reality rests on the interdependence of matter or sensibility—the inferred objectivity of the external—with mentality, or with the subjectivity of mind-internal. Mind cannot create external objects without sense-data, and loses objects when deprived of sense-data, as in snow blindness. Leaving aside non-veridical psychic images or the dissolution of space in pathology, this points to the mind-dependence of space, but confirms the capacity—obvious in dream—to generate pre-objects (images) in the absence of sensory data.

The passage from inner to outer—from the generic to the specific—unravels the simultaneity of a subjective transit into a serial progression, i.e. the spatiotemporal relations of an objective order. The temporal order of events in the mind is no doubt parasitic on objective data but, in an act of cognition, it is preliminary to temporal order in the world. The generality of the preliminary, the progression from inclusiveness to specificity, and the sequence of psychic antecedents, as disclosed in case study, are all conveyed into object space in a continuous passage from subjective to objective.

III
Acknowledgements

I would like to acknowledge the following sources: Chapter 3 is in Pachalska, M. and Kropotov, J. (eds.) (2014) *Psychology, Neuropsychology and Neurophysiology: Studies in Microgenetic Theory: In honor of Jason W. Brown*, Frankfurt/Lancaster: Ontos Verlag, in preparation. Chapter 4 is in *Process Studies*, 41, pp. 21–41. Chapter 5 is in *Journal of Mind and Behavior*, 35. Chapters 6 and 7 were both originally published in *Mind and Matter*, 10, pp. 47–73; and 11, pp. 1–21. Chapter 8 is in Weber, M. and Berne, V. (eds.) (2014) *Chromatikon X. Annales de la philosophie en procès (Yearbook of Philosophy in Process)*, Louvain-la-Neuve: Éditions Chromatika. I would also mention helpful comments on various chapters—supportive and contentious—from Harald Atmanspacher, David Bradford, Jonathan Bricklin, Gary Goldberg, Marcel Kinsbourne, Maria Pachalska, Raymond Russ, Bogdan Rusu, John Smythies, and Michel Weber. A special thanks to Maria Pachalska for helping to edit and format the manuscript and for most of the figures.

Chapter One

The Mind/Brain State: Nature

Even when the style of an absolute idealist seems most troubled by obscurity and confusion, I am often subject to an uneasy feeling that this is partly due to my own failure to see something well worth seeing which he sees dimly and so describes obscurely, but which his critics do not see at all.
—Ewing (1933)

Introduction

Theoretical writings on mind and brain tend to assume reduction or identity, except for consciousness, which is problematic, and not readily explained in terms of causal brain process. I have argued that *consciousness is a relation from proximal to distal segments in the mind/brain state*, with the segments and their relations the focus of inquiry, not consciousness as a mental function or activity (Brown, 2012). Fundamental is the nature of the mind/brain state,[1] a theory of which was advanced in prior papers (e.g. Brown, 2010; 2014). Based on clinical data that go back to articles compiled in Brown (1988), microgenetic theory posits a subjective world elaborated in the mind/brain state, an extra-personal space continuous with dream and hallucination. The position is a mode of idealism with perceptual space conceived as the outer rim of mind constrained by sensation to model the physically real, yet wholly realized in the observer's mind/brain. For most, an internal subjective sphere is distinct from an external objective one, the

---

[1] The term mind/brain state implies a unity of brain state and mental state, but is non-committal as to the relation of mind and brain. It is implicit, however, that a mental state does not occur without a brain state (as defined in this chapter), but brain events need not be accompanied by (cause, instantiate) mental events unless the events occur in the context of a brain state (cf. below).

latter conceived as mind-independent yet comprehended by an individual directly or by inference, though from a microgenetic perspective, inner and outer are subjective points of view with physical reality sampled *indirectly* in perception, outside immediate knowledge.

One line of thought separates mind from brain altogether, not only in the different forms of dualism or mind-dust theory, but in phenomena that appear to defy interpretation in terms of causal brain function, e.g. intuition, dream, unconscious thought, premonition, synchronicity, noetic and archetypal theory (Kelly, 2007; Atmanspacher, 2012). The inability to specify the relation of mind to brain, and the distinction of mental and physical facts, e.g. the *res cogitans* and *res extensa* of Descartes, justify the claims of some thinkers that the brain-independence of the mind runs parallel to the mind-independence of the world.

Symptom-Formation and the Mind/Brain State

The relation of mind and brain in different philosophies, e.g. monism, dualism, idealism, epiphenomenalism, identity and other forms of physicalism or materialism, tend to dance around the problem avoiding the effects of brain lesion on mind, except for findings that reinforce a particular claim, with more attention given to computational models and mathematical simulations than clinical data. However, errors are lawful or regular within categories, such that perturbations of brain are assumed to be comparable to the relation of the same regions or systems to normal behavior. The relation of symptom to lesion, e.g. the "wrong" word with damage to language areas, illustrates the relation of some aspect of language to that area. But, the attempt at correlation is controversial since it is not possible to replicate symptoms across individuals, and pathologies differ in kind and location. Variables such as age, gender and handedness confound localization and type of symptom. At the very least, pathology confirms that mind is not brain-independent; indeed, that specific functions of mind are associated with specific systems in brain. The fact that brain damage leads to a *category of errors* associated with some portion or activity of brain confirms that mind depends on brain even if the exact nature of this dependence is uncertain.

The usual assumption is that damage degrades or destroys a function—a representation, strategy, operation—or inhibits or releases (disinhibits) other brain areas connected by pathways. The effect can be structural, with interruption of cells and pathways, or neurochemical with a local or referred alteration. The concept of regression to earlier-acquired behaviors, i.e. that pathology recovers the acquisitional sequence, has fallen out of favor though it can provide an important perspective. The main difficulty is that a reduction to physical substrate — cell, network, chemical, gene — without an explanatory within- or cross-level strategy, fails to give a theory of symptoms in a psychological context, much less the deeper relation of mental phenomena to brain.

In prior writings, I have approached this problem in terms of the growth processes of parcellation (Ebbeson, 1984) and neoteny (Gould, 1977). The idea was that focal brain damage induces selective retardation (Brown, 1994), such that earlier phases in the mind/brain state are not transformed to ensuing phases but survive as errors in the final performance. A focal lesion does not destroy function but delays transmission, and so exposes that segment of cognition mediated by the damaged phase. This concept has been discussed in some detail with illustrations from clinical study (Brown, 1994; Brown and Pachalska, 2003). Here, the aim is to expand the idea to a more general formulation, relating errors to normal psychology and brain process.

If an error such as hallucination in place of an object, or word substitution, refers to phases in process prior to a normal outcome, the normal process would consist of phases in the brain state that incorporate the neotenous segment. More precisely, *local retardation of brain process, incomplete specification and attenuation of cognition occur as a unitary event*. The error carries through to the final outcome, e.g. a lexical error undergoes normal phonological processing. This confirms serial transmission in which the erroneous word "calls up" ensuing phases, i.e. the delay does not truncate the process but carries the segmental change through to the end-stage.

For example, disruption at a phase of lexical-selection that results in naming a table as a chair reflects retardation (neoteny) of the antecedent category, e.g. furniture, with incomplete specification (parcellation) of the correct item, sampling the background category, giving a word closely related in meaning. However, the erroneous word undergoes normal subsequent phonological processing. The segment of an error that is thrust into the foreground, e.g. a zeroing-in on the relative equivalence of chair and table within the category, and the sampling of the category, are concealed because the virtual item is transformed to a later phase. Neoteny and parcellation are patterns of epigenetic growth, essentially whole–part or category–item transitions that continue after a relatively stable structure is achieved. The pattern of cognitive process continues into adult life as an extension of early patterns in brain development (see Pribram, 1991, on embryological "force lines").

A further implication is that segments of preliminary cognition are ingredient in actual outcomes, e.g. lexical-semantics and object-concepts are not maturational add-ons but participate in acts of thought and perception. The meaning allotted to objects or words, their conceptual and affective relations, plus the wider knowledge base and/or experiential core, are not associations to a final object, but are woven into the fabric of its momentary existence. Abetted by the epochal nature of the state, in which the succession of before and after is simultaneous until one cycle of existence is complete (Brown, 2010), early phases in the mind/brain state are concurrent with the later ones to which they give rise. The study of errors with brain damage supports the idea that a mental state incorporates the entirety of its phases.

The pattern of change that underlies the process of symptom formation is identical to process in the brain state. By exposing the microtemporal sequence concealed in surface outcomes, the error is a bridge from mind to brain, revealing a series of whole-part shifts. The brain process that underlies the individuation of objects, and the "mental process" that is uncovered by clinical study, fail to undergo normal specification. The error is a maladaptive outcome, a neotenous whole-part analysis. The same process characterizes each aspect of the mind/brain state, with complementariness of mind and brain from the onset of the state to its actualization. The evolutionary progression from archaic to recent corresponds to the shift, in *morphogenesis*, from exuberance and redundancy of cells and connections to final specificity of innervation. The *cognitive process* extends this pattern in a trajectory from generality at onset to definiteness at termination. *In sum, a patterned sequence of brain activity corresponds to a patterned sequence of mental activity over the full processing sequence. This activity – a sustained whole-part transition[2] – begins in early brain structures and carries through to a neocortical endpoint.*

Microgenetic process resembles the gradual unfolding of an organic system, such as Goethe's description of a flower growing out of a seed as a kind of qualitative fractal.[3] More generally, parcellation and heterochrony (neoteny) are a source of evolutionary variation. The progression from generality to precision (Whitehead, 1933) corresponds with the transition to diversity in the specification of categories to

---

[2] Accounts of context–item, surround–center, frame–content or ground–figure transition in prior writings, or the description of individuation, specification or differentiation, or the passage from generality to precision, all allude to the whole–part shift detailed in microgenetic theory.

[3] Microgenesis has been compared to fractal theory (MacLean, 1991; Vandervert, 1990), though the fractals are qualitatively unique, not self-similar replications. See Pachalska and Weber (2008) for a recent compilation of papers on microgenesis and research data in support of the theory (especially papers by Don Tucker and Talis Bachmann).

particulars. The analogy of mental and brain process in microgenesis recalls speculations by some Gestalt psychologists (Köhler, 1938) on isomorphism, but they invoked geometric patterns of brain activity corresponding to perception while in microgenesis the isomorphism is the common pattern of individuation in mind and brain.

A selective retardation in development not only explains *errors* in cognition and their relation to normal function but also *advances* in brain morphology (Gould, 1977) and the *creative* growth of thought (Brown, 1994; Brown and Pachalska, 2003). An example is the rapid expansion of the human neocortex, in which cessation of brain growth is postponed so expansion can continue into post-natal life. This is accompanied by a delay in closure of the cranial sutures, which is also a neotenous feature. In cognition, neoteny accounts for the growth and propagation of thought at unconscious phases.

If we extrapolate from early development to pathology in maturity, we can explain the elicitation of novel antecedents by a prolongation that allows for: (1) the surfacing of submerged phases; (2) the intrusion of preliminary cognition as a conscious endpoint; and (3) errors as off-target outcomes of the affected category. In a word substitution such as saying chair instead of table, a lexical category (furniture) is sampled prior to the individuation of the intended target. Hallucination realizes a normally traversed segment of imagery that is exposed in the neotenous segment. Similarly, thought is an elaboration of the imaginal underpinnings (inner speech) of language (Vygotsky, 1962) with a delay in the passage to vocalization.

The point is that growth in morphology and growth in function occur with prolongation of juvenile or preliminary stages. The juvenile in morphogenesis is equivalent to the preliminary or archaic in brain and cognition. Novelty in thought, advance in evolution, whether adaptive or anomalous, do not occur at the end-stage of development but at earlier less specialized stages. In pathology, when automatic transition is delayed, the phase expands and maladaptive contents are elicited from a relative stasis (recurrence) of the category instead of individuating an appropriate item. Creativity entails a proliferation at phases that are ordinarily unconscious. Novelties are carried-through to consciousness. One could say, again citing Goethe, that the individual returns to the pool of the creative unconscious, but a more accurate description is that the dominant focus of cognition is at segments of choice, introspection, branching, propagation and deviation from expectancy. Creative ideation occasioned by the neotenous prolongation of sub-surface phases permits early cognition, e.g. metaphoric, syncretic, paralogical, imagistic, to displace adaptive outcomes in speech, rational cognition and object perception.

The combination of slowing or retardation, and extension of the juvenile into the mature, with specification, parcellation and sculpting of stages of greater generality, imply that cognition is a continuation of epigenetic algorithms that mediate the transition of genome to phenotype (e.g. Katz, 1983; further discussion in Brown, 1994). *In sum, the elimination of cells and connections in morphogenesis, the physiological inhibition of alternatives after some degree of stability is achieved, and the context–item transformations in the actualization of the mature mind/brain state, represent a continuous whole–part specification in which trends that are established in fetal growth continue into maturity as the cognitive (microgenetic) process.*

Comment on embryogenesis and the growth of mind

The pattern of fetal growth as the model of cognitive process can, theoretically, be traced to the still earlier pattern of mitosis, or cell division within a membrane from the fertilized ovum to that of mature cells. This is individuation from within of parts out of a whole, with specification from unity to multiplicity, not the reverse. The elaboration of fetal cells within an outer membrane is an internal specification. This leads to increasingly specialized subsystems — like a plant evolving from a seed — that develop to such an extent their relation to the originating (stem) cells can no longer be discerned. In mind there is also progressive specification to seemingly autonomous and interconnected units. Yet regardless of final complexity, the basic pattern is progressive individuation from within by way of internal relations, not an association or addition by external relations.

The eventual splitting (partition) of the infant from its mother in birth (parturition) completes one physiological cycle, just as another — psychological — one begins. Like the sperm and egg that fuse to a unity, newborn and mother fuse to a single entity (from the infant's perspective). The incipient mind of the infant incorporates the immediate outer field, e.g. the breast. Gradually, the breast (mother) separates and an outer field develops in opposition to the inner one. However, the outer field — the proximate world of the infant — is the externalized portion of the infant's mind. That is, a segment of the infant's subjective field objectifies as the outer rim of its mind. The infant is in relation to — and cognition goes in a direction to — the externalized portion of its own subjective field. The intrapersonal segment develops largely along intrinsic lines; the extra-personal segment is honed by sensation to model the material world. Mind grows by recurrence as process develops from inner to outer. There is incessant replacement of the mind/brain state with the outer field realized from the inner one, each iteration serving to consolidate the binary organization. The splitting of outer from inner repeats the splitting of cells in mitosis, then infant and mother, but in the psyche the outer is part of the inner, just as mitotic division occurs within a cell membrane.

Within the subjective portion of the infant's mind, a core self develops in concert with the expansion of the outer field and the increasing articulation of external objects. The objective field is an objectification of the subjective field. The confrontation of subject and object, or mind and world, is an opposition of an inner segment of subjective mind with an outer segment of subjective mind that has externalized and "detached" by way of sensory constraints. The stage of subject–object relations prior to a self-concept is probably the final achievement in most animals, but in the child it is an initial phase in the growth of the self.

A comparison of cognition to germinal tissue is more speculative, but the analogy is compelling. I have previously noted that the outer world is like the skin of the mind. To see this analogy we have to understand that object-formation occurs over a sequence of phases from an unconscious core to the outer world. The same development occurs in all cognitive domains, namely a rapid migration from an onset in older regions of brain evolution to areas of greater recency. The neocortex is an endpoint in brain evolution and mental process. The succession of phases in object-formation from potential to actual, to perishing and replacement, is analogous to, or has a deep commonalty with, patterns in the embryology and histology of the skin.

Brain and skin are derived from primitive ectoderm. The neocortex is laminated like the skin, and in mental process, as in epidermis, there is a continual replacement of objects that are born, grow and die. The epidermis consists of five layers. The deepest layer, the *stratum basale*, contains cells that undergo mitosis, migrate to the surface and are shed. There is continual replacement of cells on the external surface of the skin by those generated out of deeper layers. The migration of skin cells to the surface where they die and are replaced is analogous to the generation of objects over phases. Like skin cells, final objects externalize and perish as they are being replaced.

In sum, microgenetic theory opens the door to a rethinking of the relation of cognition to genetic process in a coherent account of the growth and evolution of mind and brain. Basic patterns in evolution and morphogenesis, in cell division, replication and growth, correspond with patterns in the momentary realization of the mind/brain state. The implication is that the foundational laws or principles that govern the growth and reproduction of organisms also govern the process of thought, action and language. This is clearly a topic worthy of further study.

Toward a Theory of Mind and Brain

The analogy of process in mind and brain states points to an iterated specification through multiple partitions over a hierarchy of *mental* categories or *neural* configurations, with the specification of configurations in the brain state accompanied by the specification of categorical wholes in the mental state. For example, drive-categories specify conceptual-feelings, which lead to images — thoughts, memories — and finally, to acts, words and objects. This entails that brain does not cause mind, which requires a causal or temporal step from brain physiology to psyche, but is similar or identical to it. In a word, a description of the process underlying the brain state is a description of the process underlying the mental state (Brown, 2014).

A consequence of this theory is that brain activity that is incoherent or disorganized would not generate a conventional mental state, or the state would not resemble the ordinary concept of what a mental state is. Mental activity that is not characterized by feeling, i.e. not emotion but feeling as the driving force of organism, or activity that lacks a pattern of whole–part transition, would not generate a mental state.[4] The mental states of animals tend to be described according to their approximation to the human model. When such states fall on the main evolutionary line, even with wide deviations, e.g. sonar, animal-mind is still no closer to machine-mind in spite of scant evidence of inner states in all but the higher forms. Mental states in animals show a similar albeit less developed pattern compared to human mentality, though it is difficult to apply to animals the usual criteria of subjectivity (cf. Griffin, 2001). However, even in the lowliest unicellular organism, tropisms and the vectors of approach and avoidance represent the internal activity of organism, directedness and the role of feeling in behavior.

Can one say process in the brain state is necessarily accompanied by a mental state or, conversely, that a mental state is necessarily accompanied by a brain state? We tend to think the brain state generates the mental state, not the reverse. It is plausible to say a brain state characterized by a multi-tiered series of whole/part shifts is necessarily accompanied by a mental state, i.e. that the brain state is not independent of the mental state and *vice versa*. It would seem inconceivable that a mental state could occur without a corresponding brain state. Brain state and mental state cannot exist without the other. While a specific pattern of brain activity is necessary to generate a mental

---

[4] In philosophical writings, an image, a proposition, a *quale*, is often taken as constituting a mental state. In microgenesis, these contents do not stand alone but are segments with a diachronic history.

state, and while random brain activity presumably does not give a mental state, as in a seizure discharge, the activity of even one neuron is probably not without some proto-psychic features. If mind traces to rudimentary entities, any brain activity or component would manifest some precursor of psyche, e.g. feeling.

Microgenesis entails some version of panpsychism (Weber and Weekes, 2009; Brown, in press) with activity in primitive entities describable in terms of precursors that evolve to psychic states. Beginning with elementary particles, the physical realizes some proto-psychic features. Within the physical sphere there is an evolutionary sequence from simple to complex organisms, though there is disagreement as to the onset of proto-mind, i.e. the requisite pattern of brain activity, its development, continuity and replacement, as well as gradualism versus emergence. If physical entities evoke proto-mind, could proto-mind evoke physical entities? Can psyche exist independent of physical entities, for example the postulation by Eccles (1992) of "psychons" attached to "dendrons", Hartshorne's (1962) ideas along the same lines, or the mind-dust theory critiqued by James (1890). These speculations involve free psychic elements concentrated in or by the brain.

If every physical entity has some proto-psychic features, could psychic entities have proto-physical features? If so, how would they be characterized? On this view, nature is a precipitate of cosmic mind. Could mind evolve without neural correlates to the point where, because of complexity or sustained development, physical process and, ultimately, the brain are enlisted? Is psyche part of nature or distinct from it? Many believe that mind can exist independent of nature or physical substrate, as in spirits, souls or ghosts. Yet if a mental state corresponds to a brain state, and the reverse, brain states are not independent of mental states and mental states are not independent of brain states. If the physical is bound to the mental all the way down, the mental is bound to the physical all the way up.

The implication is that brain is no more independent of mind than mind is independent of brain which, if true, rules out the possibility of brain-independent mind. It also eliminates the possibility of a mind that is independent of other physical substrates even if every physical entity has proto-psychic features, since a mental state requires a *specific pattern* of physical activity. A computer that does not realize a hierarchic succession of whole/part shifts as in the brain state would be incapable of generating, or accurately simulating, a mental state, though the physical components of a computer, even with the power turned off and without cooperative interaction, might, individually if not cooperatively, have some primitive quality of proto-mentality, assuming that the antecedents of mind inhabit basic physical entities.

On this view, machine intelligence, though imitating some features of human mind, and perhaps exhibiting some precursor of mentality in its constituents, does not reproduce the mental states of organic systems; it is, at best, a model of some output manifestations, not the internal process behind them.

This analysis depends on the definition of mind and mental state. I have given a fairly detailed account of the mind/brain state. On the other hand, as Wittgenstein famously noted, philosophers who speak of mental states tend to leave their nature undecided. If mind is problem-solving, rote memory or logical reasoning, one could argue for computational mind. But if mind evolves from feeling to drive, desire, reflection, imagination and dream in a momentary recurrence of epochs, with a distinction of inner and outer, a revival of past into present and a felt subjectivity, there is little comparison.

Identity and Monism

The conflation and common patterns of mental and brain states, if documented, provide important insights to a theory of mind–brain relations. Clinical evidence supports an approach in which a change in the brain state is a change in the mental state, while any change in the mental state, however subtle, implies a change in the brain state. The evidence that supports this idea, which originally led to microgenetic theory, can also be used to refute it. For example, one can have large lesions of the brain with minimal or non-detectible mental changes. Nor is it clear that every mental state, including those of psychopathology, corresponds to an alteration in the brain state, but this depends on the *content* of the state, not the process of its realization. Generally, a change in the mental state is adduced from a change in its *content*, while in terms of *process*, the pattern of the state is primary, not the momentary details of what the state is about.

It does seem obvious that saying or thinking the word *chair* instead of *table* corresponds to a different mind/brain state, minimally in the configuration that corresponds to the different words. The mind/brain state differs from one moment to the next and is probably never the same across moments and individuals. For someone to say that different brains share the same mental state because they share the same idea or proposition limits the mental state to the content at a given moment irrespective of the experiential history of the individual, the momentary becoming of the state and its spatiotemporal context.

As to mind/brain relations, for microgenesis, mental and physical entities—minds, brains, particles—are embodiments of internal relations. The mental state is not a causal effect or output of the brain nor a product of brain activity, but develops intrinsically as an aspect of the brain state. Since mental states and brain states refer to the same state, the identity of pattern implies that one does not cause the other. The progression from one phase to the next within a state is a specification of potential, not a causal sequence, though the replacement of states in those portions that overlap may involve causal relations. *The theory holds that early phases do not cause but become later phases, like upstream segments in a river become downstream segments, or the jet of water in a fountain becomes the spray at the surface* (Brown, in press). This is more like causal persistence (Russell, 1948) than ordinary causation, i.e. a thing replacing itself, as in the subtle moment-to-moment changes of maturation or aging. States, and the continuum of phases within states, are jointly specified. We tend to accept that brains are essential for minds, or that the mental state is dependent on the brain state, though we do not ordinarily believe the reverse is possible, i.e. that mental states cause brain states, or that the brain state is dependent on, or an effect of, the mental state.

Many people believe, and probably all people feel, that mind has agency and that actions are causally implemented (Brown, 2012). We suppose that consciousness must have a function, must do something to fulfill the purpose for which it evolved, so it is not implausible to argue that mind influences brain, or that a self or consciousness has a causal effect on behavior, thus on the brain. We believe that a self can judge, decide and act, i.e. agent causation. This is an implicit recognition of the powerful impression that mind acts on brain to cause physical effects. A causal role of the self—consciousness, decision, reason—is part of the larger problem of the relation of mind to brain, whether brain causes mental activity, or mind causes brain activity. The self as a content in the mind is part of the overall concept of mind as a content in brain. We assume that brain activity can occur without mental activity, e.g. discharge in an isolated slab of brain cells, but we have difficulty conceiving mental activity without brain activity or that a form of mentality might accompany discharge in the slab. While in theory a causal relation might apply in either direction, the belief that brain activity is primary has stronger appeal. There is much evidence that mind depends on brain, e.g. the attrition of mind with brain disease or damage, and no clear evidence that mind can exist independently.

Arguably, an occurrent brain state has a causal effect on an ensuing brain state, but does a mental state have a causal effect on the next mental state? Though we assume that brain activity at one moment causes, or influences, brain activity at another (the following) moment, it is unclear how *mental* activity could cause other mental activity independent of brain. If mind has a causal effect on brain, it must possess some property that is responsible for the effect. What would this non-physical property be? Further, if mind has an effect on brain, it should be capable of an effect on other mental states, presumably by way of the same causal properties. If mind has an effect on other mental states independent of brain, why should those states be restricted to the individual's own brain or mind? A causal mental state opens the door to mind-reading, telepathy and a variety of controversial phenomena, including the survival of mind after death. However, what is conceivable is not necessarily what is plausible, and what is plausible is not necessarily what is possible. It is difficult to see a path into the idea of a causal role of mind or consciousness that has a modicum of convincing evidence.

Content in mind is not an isolate but is part of a micro-temporal sequence. Mind-to-mind causation, or the causal relation of mental states, is confounded by the presumption of causation for the acts and objects that are the externalized endpoints of the mental state. That is, object-causation—the perceived effect of one object on another—is, in fact, a relation of one mental state to another. The outcome of a mental state—an act or object—is an image or object-presentation[5] that corresponds with—adapts to—physical entities in the world. An object perception, by way of sculpting to an adaptive fit, is a mirror of macro-physical conditions in the external world, but it is nonetheless a mental image. An assumption of causation among physical entities does not readily translate to an assumption of causation among mental objects, only that objects replicate the appearance of causation in some way other than by causal relations. Since mental images (objects) do not have causal properties, causation is simulated by the immediate recurrence of states. Change is not transmitted from one object to another but is, for the observer, an illusion of changed replicates.

The appearance of a causal relation between objects is anticipated by the feeling of agent-causation in the relation of thought (mental imagery) to acts, which are externalized thoughts. Put differently, the causal effect of thought (Brown, in press) is an attribution of causation to an earlier phase in object-formation, while object-causation is an attribution of causation to a late or distal phase in the mental state. The felt agency for acts and the inference of causation for objects is transferred to—more likely arises from—their immediate precursors in the mental state. Object-causation is the interpretation given to the replacement of mental states, not the causal relation across the perceived end-

---

[5] In the sense of an immediate awareness of an object that is presumed (in my view, falsely) to be independent of belief and judgment, i.e. mind (discussion in Stout, 1902).

points of the states. Even if mental causation occurs, it would not be "top-top" (object to object), or from one conscious state to another, but "bottom-up" (replacement of states) over the entire transition (Fig. 1.1).

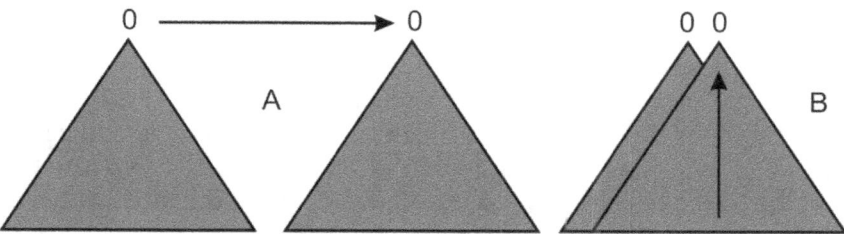

In Fig. 1.1: A, the ordinary concept of object-causation, has the endpoint of one mental state (a perceptual object) as the cause of the endpoint of another mental state. For microgenetic theory, objects (0) are the final actualities of mental states that enfold a diachronic history (B), with overlapping replacement giving the appearance of causal passage as an adaption of mind to reality.

If it is reasonable to ask how a mental state without obvious physical properties could *cause* a change in material brain process, it is also reasonable to ask how brain process could cause mental activity. What properties or features of mind could be causal objects; i.e. how would a material brain act on immaterial mind, and the reverse? An absence of a causal relation from mind to brain would seem to entail an absence of causation from brain to mind. This is consistent with the thesis that mind and brain are different "stuff", but it is also consistent with the idea that mind and brain are *dual aspects* of a common process. Microgenesis entails internal relations and the primacy of the subjective, i.e. a mental state intrinsic to brain and *vice versa*, with mind and brain exhibiting the same pattern of whole–part derivations.

Whole–part transition and dual-aspect
Generally, the dual in dual-aspect presumes a correlation of mental activity, especially consciousness, with physiology and neurochemistry, with the expectation that one can go from mind or consciousness to neuroscience without passing through the infrastructure of experience. The mental state cannot be contracted to consciousness without incorporating the sequence of phases out of which consciousness develops, and brain activity cannot be reduced to chemical reactions without relating the activity to the hierarchic structure and evolutionary process of brain activity. A dual-aspect approach has consequences similar to reduction or identity though it is not a materialist theory. The mind is an aspect of brain activity, as brain activity is an aspect of mind. This entails complementary aspects, with both aspects

arising from a source that is neither mental nor physical, or both mental and physical. The theory is coherent and self-sufficient, and avoids dualism, but it does raise the possibility of a neutral monism in which mind and brain refer to, or are realizations of, a deeper reality.

The deeper reality to which monism refers might be all-mind or all-nature. On the *first* possibility, the mind/brain would be a realization of pure subjectivity, an ephemeral moment in an eternal *Idea* like a thought in the mind of god. If the passage from the Absolute to primitive organism or to the antecedents of mind/brain is abbreviated in purely mental entities, of which we have no good examples, the ultimate foundation of matter would be *Idea*, and matter would be a concept in individual mind, which itself is a concept in all-mind or the mind of god (Inge, 1924). Philosophers have alluded to this possibility, notably Berkeley and Schopenhauer. This theory might appeal to the absolute idealist, for whom all experience and the reality behind it are phenomenal.

The *second* possibility is that the mind/brain is a realization of non-psychic nature, typically, that mind develops at some stage of complexity in the evolution of organic systems. Compatible with both accounts is the concept of the brain state as an *appearance* in individual mind. We know or surmise that the brain is a perceptual object in the mental state, a phenomenal appearance as are all objects to which physical existence is imputed. All life, all experience, all thought, all physics, are filtered through the mind. Brain activity attributed to physical (physiological) process is still an event in perception. The physical brain in reality is inaccessible, though a phenomenal brain is presumed to be the cause or correlate of a content in phenomenal mind.

If the mental state is intrinsic to brain, and not independent, parallel or epiphenomenal, a *third* and more likely possibility is an origin in potentiality with proto-physical and proto-psychic features that develops to material stuff with an aspect of subjectivity and mental stuff with an aspect of physicality. This deeper reality (Absolute) from which all phenomena arise constitutes the fundamental "ground" of matter and life. An Absolute would evolve to organism as category and feeling take on directionality. The question is, at what point should this process be assigned a reality that underlies phenomenal experience?

We can reduce mind to brain, brain to gene, gene to molecule and so on, in fact we can reduce all entities and objects to energy. But this reduction does not explain at a lower level what is left unresolved at the higher level, nor does it contribute to the translation of lower to higher. For example, to identify the genetic code that makes language possible is a far cry from explaining how we say the word *chair*. To identify a brain area involved in speech is not to explain the function of that area in speaking, though many who are ignorant of the specifics still debate the generalities. For (neutral) monism the critical step is the transition to feeling, with a shift to directionality as the nucleus of a subjective aim.

It is well enough to speculate that energy (strings, vibrations) is the ultimate source of all things but further explication is needed for the evolution of energy to feeling and the growth of feeling to the manifestations of organism, *viz.* category or potential and its partition, directionality of feeling and subjective aim. Monism is compatible with dual-aspect in the passage from absolute reality to the unity of a primordial physical/psychic entity, but the postulation of a substance or relation that is the absolute bottom of the phenomenal world does not add greatly to the explanatory power of dual-aspect, though it does raise questions on the evolution of a higher mentality and the common origin and development of the mind/brain. Of import is whether the concept of whole-part transition and the mind/brain state, which is the outcome of theory based on clinical study, and which I have argued are critical to understanding mind/brain process, can be conceived as fundamental to material states of energy before their derivation to organic systems.

Neutral monism would seem to require panpsychism all the way down to the ultimate foundations of being. If the passage of this Absolute to physical entities realizes a stage prior to evidence of mentality, such as a rock, the foundational process of life and matter and the ultimate source of being would be relegated to a physics of non-subjectivity. For most, a rock is an aggregate devoid of subjectivity. For Whitehead (1954ed), it is a "mass of raging particles". Does the orbiting electron in an atom have a direction that is compatible with a subjective aim? If so, is it uniquely physical?

To me, the most plausible assumption is that the mental and the physical are inseparable from the onset. The difficulty is that the evolution of organic life has occurred over millions, if not billions, of years with mind and subjectivity occurring at a relatively late stage and in complex systems. Since there is no reason why a strictly physical entity should evolve to mental states, or even progress to subjectivity in simple organisms — why not an earth barren like the moon? — the evolution of mind is unlikely to be purely contingent on, or wholly shaped by, a changing environment, as important as this is, but would owe to proto-mentality at an elemental stage of physical matter, i.e. a pre-psychic complement that establishes a subjective aim and directional feeling. Directionality would provide a forward thrust to evolution that, constrained and facilitated by the environment, propels it to further levels of development.

Chapter Two

The Mind/Brain State: Origins

> ...the facile propensity for fleeing to "nature" as
> a city of refuge for theories in distress.
> —Laird (1929, p. 197)

Synchrony of Act and Object

Many neuroscientists over the years, from Pavlov to Herrick (1924) and Sherrington (1951ed), and most recent authors, maintain that cerebral function and the higher cognition are mediated by a compounding of reflex to increasingly more complex organizations, an approach that derives from the synaptic model of nerve function (Cajal, 1954ed) and is consistent with the presumption of "all or nothing" discharge, excitation/inhibition, causal theory of brain activity, and digital or on/off computational models. The reflex account permeates all disciplines of mind/brain study, from the analysis of individual cells to the study of large neuronal populations, as well as theories of cognition.

The box-arrow or flow-models of cognitive science are a product of this concept as is brain study in neuronal connectivity. The model has survived several indirect assaults since Cajal's attack of the syncytial theory, notably the concept of emergence and oscillatory systems (Dewan, 1976), Gestalt theory (Wertheimer, 1945), Lashley's mass action (1964ed) and the holographic model of his student, Karl Pribram (1991). Even the modest suggestion by a leading researcher of synaptic function (Eccles, 1970), that cortical activity might be better understood in terms of wave fronts or field effects, has had little impact. The famous review of Skinner's model of language by Chomsky only substituted for input and output to the "big black box" of behaviorism, the smaller boxes and arrows of cognitive "science". These proposals, along with related theories in philosophy, e.g. connectionism, modularity, in spite of their pretentions, are little more than refinements of the theoretical target of Chomsky's critique.

An approach to brain study that begins with an account of spinal reflexes leads to the postulation of more complex systems in a passage to higher levels of brain function. Theories of assembly or construction in perception are consistent with the reflex model in that they are built on the idea of external relations among logical solids, which follows the natural tendency to reduce problems to lower level explanations and consider the lower levels as building blocks of the higher ones, instead of exploring whether the latter can give unexpected insights to the evolution and nature of the former.

In my writings, the transition from reflex to representation, uncovered in the study of breakdown patterns of language and human cognition, and published some years ago, was again described in Brown (2010), along with physiological evidence for this hypothesis. The aim here is to review the hypothesis briefly and, with the benefit of subsequent study and reflection, discuss the implications of this transition, specifically, the shift to field effects from interaction and point-to-point transmission. Accounts of the evolution of human cognition tend to assume increasing complexity of reflex design with little attempt to examine the nature of the complexity. Presumably, at some point the complexity is sufficient for the emergence of consciousness and introspective content, but the process(es) involved in the complexity, apart from greater numbers of cells and connections, is left undecided. This includes the most basic questions, such as the nature of the mind/brain state and the development and origins of action and perception.

The microgenetic account of increasing complexity supposes that the interval between self and action, which is the conceptual space of introspection, undergoes an accentuation, with an expansion by the partition of categories in perception and language prior to externalization. The unified and parallel unfolding—*becoming*—of action and perception entails a cascade of whole–part shifts that correspond to a qualitative fractal-like specification at multiple levels. The partition generates complexity from within, not as an addition to pre-existing systems, by way of a serial emergence of configural elements nested in background categories. Indeed, the unpacking of categories resembles in some respects the original critique of the behaviorist model of language by the illustration of embedding or recursion in the relation of parts or members to categories, e.g. the embedding in a sentence of subordinate clauses.

According to this hypothesis, the evolution of the human brain involves a progression from the circularity of a stimulus/response chain, i.e. the closed loop of reflex, in which each cycle is repeated with little change (Weiszacker, 1939/1958) *to the synchronous derivation of an endogenous act/object.* In reflex the response to identical stimuli is relatively invariant, and can serve as a stimulus for another round of reflex activity. This is characteristic of simpler organisms and many processes in the spinal cord and human brain, especially the brainstem, as in respiration and swallowing. The simplest level involves a direct or mediated sensori-motor connection. The next stage, in brainstem, entails a proliferation of small internuncial cells between the input–output or sensory–motor limbs of the reflex arc. The proliferation expands to a population of small cells that generate a *unified construct* that avoids the physical inevitability of reflex, with the onset of an endogenous, subjective process and the possibility of creative advance. The initial construct diverges in parallel tracks over hierarchic levels in forebrain evolution. This combined action/perception forms the basic plan of the mind/brain, and becomes a configural wave that develops over a psychic infrastructure lodged between physical tiers of sensory and motor keyboards (Fig. 2.1).[1]

The shift from *physical sensation and movement in reflex, which are external to mind, to the incipient subjectivity of action and perception, which are internal and psychic, lays down the mind/brain state*. Perception and action are its primary components and subjective pillars (Fig. 2.2), while sensation and movement are extra-psychic events that are, respectively, limitations and effectuations of endogenous form.[2] The act/object, organized within bodily space at upper midbrain or hypothalamus, and continuing to a neocortical endpoint, is mediated by a traveling wave that is parsed by sensation in a progressive adaptation to the outer world. At each segment, vibratory levels in action are read-off into motor keyboards — a rhythmic sequence of oscillators — that serially discharge into axial, postural, proximal and distal motility (Brown, 1988; also Bernstein, 1967).[3] The shift from nature to mind, from (physical) sensation to (mental) perception, or from (mental) action to (physical) movement is the interface of subjective experience and external reality. The onset and timing of each component are kept

---

[1] An example of representation within the input–output sequence of reflex mediated by the pre-tectal region is the "optic grasp" of a frog in which a fly is seen and caught by the tongue as a single event, an optic grasp, not a sequence from stimulus to response, or from seeing to grasping. Some examples in human brain are discussed in Brown (1988).

[2] Sensation refers to physical sensibility outside cognition; perception refers to psychic appearance ingredient in the mental state. A similar distinction holds for movement and action.

[3] See Martin (1972) for a discussion of rhythmic levels in language production.

in synch by intra- and inter-hemispheric pathways within and across widely-separated hierarchic systems.

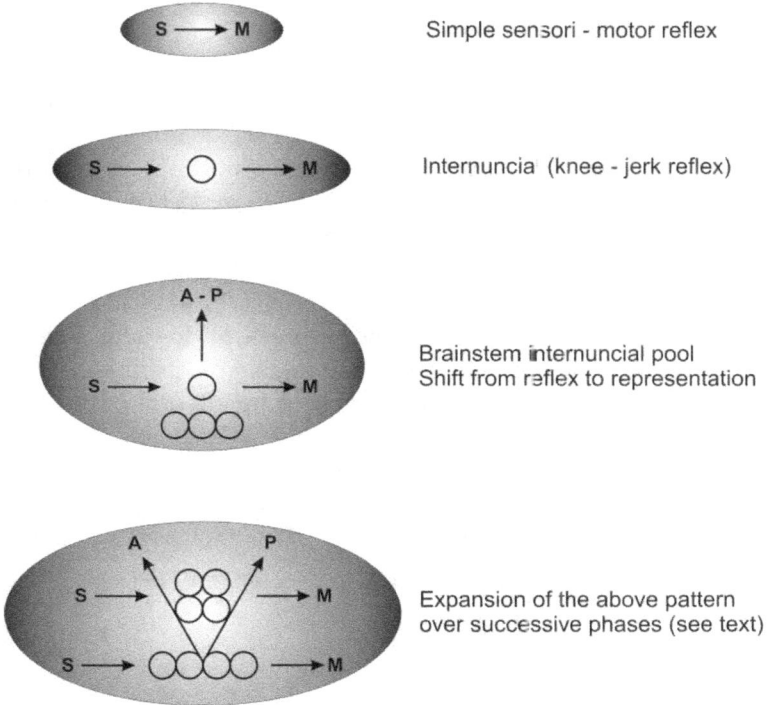

Fig. 2.1: S = sensation; M = movement; A = action; P = perception. The circularity and seriality of reflex shifts to the unity and simultaneity of representation in brainstem where small neurons interposed between the sensory and motor limbs of the reflex arc elaborate a simultaneous perception-action. The unified A/P diverges toward neocortical areas, with corresponding levels kept in-synch by intra- and inter-hemispheric pathways. Action tends to involve anterior regions; perception posterior regions, but the transition from onset to final act or object is coupled and phase-locked over the sequence.

Preliminary phases in percept-development actualize mental content, e.g. thought, while comparable phases in action contribute a feeling of self-initiation—from purposefulness to volition—essential to the distinction of active and passive movement. Action does not deposit mental content. Action is pure transition and imperceptible. Since Aristotle, motion has been related to time, which as Kant (Al-Azm, 1967) put it, "cannot by itself be perceived". Movement is secondarily perceived by collateral recurrence, i.e. a conscious action is a perception of movement. The dynamic in perception is stabilized in an object, while successive actions are seamlessly woven into events. The change from state to state, and the change within a state, i.e. motion, time, process, are inaccessible except for the feeling of activity. This feeling, the so-called *Innervationsgefühl*, was the topic of a lively debate between Wundt and James (Brown, 1988; *et seq.*; review in Bricklin, 2014).

A discussion of the origins of representation[4] is an account of stages in action and perception as realizations over a hierarchic structure. There is evidence for a micro-temporal passage in action, in studies of the readiness potential (Libet, 1985), and in the temporal lag of perception, as well as the perceptual-moment hypothesis, e.g. work by Stroud (1956) to the present. As they develop *in parallel* from a common origin in upper brainstem, the connectivity across components maintains levels in-phase as they diverge and actualize at the cortical surface. The derivation from core to terminus, which enfolds a single mind/brain state, gives rise, like branches on a tree, to successive levels in the production and comprehension of language. Other modes of cognition—emotion, memory, imagination—can be interpreted in relation to phases or segments in this derivation, pre-linguistic in animals, linguistic in humans. Thought (imagery) corresponds to pre-object phases in perception, while action deposits in bodily space—limb movement, vocalization—antecedent to the extra-psychic space of objects (Fig. 2.2).

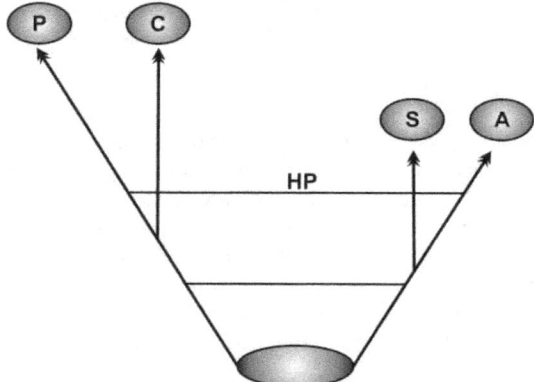

Fig.2.2: P = Perception. A = Action. C = Language Comprehension. S = Speech or Language Production. HP = Intra- and Inter-Hemispheric pathways. The primary limbs of the mind/brain state are action and perception. Perception develops to extra-psychic space; action deposits in a more preliminary bodily space. Auditory and language perception are referred to external space. Action and vocalization

---

[4] A representation is here defined as a conscious mental content, whether apprehended in the mind, e.g. a thought, or perceived in the world. The representational content is the entire infrastructure through which it develops.

are largely in bodily space. Connecting pathways (HP) keep A and P in synch at successive levels.

Transition to the Mind/Brain State

An early phase of instinct carries impulse to drive where behavior appears self-generated, e.g. hunger. This opens a window of possibility on the all-or-none nature of reflex. With instinct there is recurrence and periodicity, internal build-up and pressure for discharge, as well as some freedom from stimulation and increasing diversity in drive expression, e.g. the sexual drive and/or display from birds to higher mammals to humans.[5] An iterated reflex becomes refractory with repetition; drive becomes refractory with satisfaction. Instinct can be aroused internally and initiated voluntarily, e.g. swallowing, respiration.[6] Stimuli on or in the body are, for instinct, internal primes or external conditions. The transition is from mechanical triggers to innate patterns, from invariance and encapsulation to a response that develops over time, e.g. prey selection, hunting. Ethological accounts of displacement, consummatory behavior and satiation mark the increasing fluidity of instinct, even though it is commonly regarded as a more complex reflex. There is greater dispersion of act and object over the evolutionary sequence from machine-like behavior, as in insects and lower life forms, to complex patterns as in the mating rituals of birds. The progression to intentionality, purposefulness, individual variation and greater nuance of expression in mammalian behavior, e.g. sexual selection, territorial behavior, still falls within the parameters of the instinctual repertoire.

Reflex involves a closed circuit of stimulus and response. Representation entails a mind/brain state that is epochal, modular and reinstated. The encapsulation of reflex is for the S-R chain, while for representation it is for the entire mind/brain state. Reflex is circular and each cycle is identical. Representation is processual and each module is recurrent. The pattern of instinct is similar over instances, but details of the response can change. Hunger may lead to improvisation in choice of prey or mode of attack. An organism has a different behavior when solitary or in a group. The thematic of cycles and recurrence is retained in instinct, though unlike reflex, instinct is

---

[5] The process leads from immediate display to internal phenomena. In humans, sexuality does not externalize directly in behavior but, as with other activities, undergoes a diversification in thought and imagination (Brown, 2012).

[6] Respiration is a good example of the transition from reflex to purposeful behavior, since it can be under voluntary control, and lays down the timing for breath groups in language production.

dependent on conditions in the body and the world, i.e. the *Umwelt* of Uexküll. The coupling of act and object facilitates deviation from rigid patterns and the substitution or displacement of one instinctual expression for another, e.g. the precedence of maternal over selfish behavior. When an animal eats a prey, hides it, plays with it or feeds it to juveniles, opposing instincts compete for priority. Moreover, within each routine, behavior adapts to changing circumstances, even in the exploitation of environmental artifacts, e.g. sticks or rocks as tools or aids in the implementation of an instinctual act.

The distinction of the process underlying reflex from the process generating the mental state is meant to dispel the idea that mind is a manifold of reflex, but this should not obscure the fact that reflex is an essential precursor to a mental state. A subtle shift in design, in which a mental state arises as an expansion *within* a reflex arc, i.e. a proliferation of nested elements, becomes a pattern for further growth (Fig. 2.3).

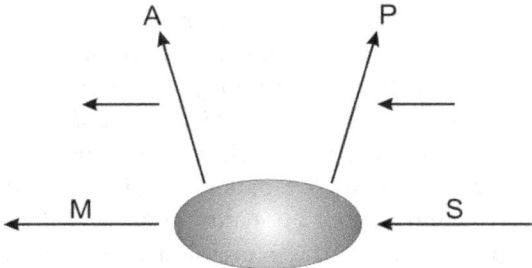

Fig. 2.3: The sensory input (S) and motor output (M) of reflex are the foundation for the parallel derivation of action (A) and perception (P) out of the cell mass nested between the limbs of reflex. The pattern is sustained at successive levels with expansion to categories, while sensation and movement remain external to the endogenous development. The concurrent derivation of action and perception creates a representation out of the internuncial pool of reflex, while the stimulus and response of reflex, which are extra-psychic events, recur as intrinsic constraints or effectuations on the infrastructure of the mind/brain state.

The combined act/object develops over phases to deposit the mind/brain state, a *becoming-into-being* that is a partition of categories. The wave-front of whole–part specification has a momentum to completion that, in microgenetic time, creates a novel present and a surge to the future, and, over evolutionary time, generates increasing complexity and diversity. Change is in the becoming (realization) of the state as a dynamic sequence, not in the being (category) in which the sequence is embedded. Deflection in passage over phases in the mind/brain state can arise as qualitative "fractals" in each partition. The state actualizes all of its phases — virtual (transformed) or actual (representations) — as "members" of a category, i.e. as the being or existence of a state (entity, organism). The entire state is a singular occasion; dynamic in actualization, static on completion. The state, e.g. act, object, idea, is the totality of what actualizes. Change occurs within the state and its replacement. Creative advance in each state depletes some quanta of potential for that occasion.

The Contact of World and Mind

The impingement of physical sensation on the endogenously-developing mental state invites a closer inspection of the nature of this contact from the standpoint of microgenetic theory. The question may fairly be asked, how we make sense of the effect on thought of the raw data of sensation. The contact of concepts with sensory data is the presumptive basis for factual statements and truth-judgments, which, so it is often noted, e.g. McDowell (1994), can neither be derived from non-conceptual physical data nor wholly grounded in an intrinsic conceptuality. The gap from the physical to the mental is construed as a transition of physical inputs to subjectivity, a difficulty finessed in cognitive science in the postulation of a causal effect of sensation on brain mechanisms which, through the appropriate computations, generate concepts and objects, an account that only reinforces the problem of relating physical data to mental states. To the extent this problem is addressed it usually takes the form of the role of experience in conforming direct perception to belief, or adapting belief to a material world. Many philosophers accept that the engagement of sensation with an inner system of concepts is causal, though some try to have it both ways, balancing a rejection of the closed circle of idealism with a boundless field of concepts or a mind-independent reality. The position requires impressions from outside to be equipped with conceptual content, but the argument, which may have merit, flounders, it seems to me, on the disclaimer that concepts are active in some experiences and not in others.

A major difficulty concerns the notion of the *taking-in* of the external world of sensory experience, which assumes that the world is not an active creation of mind, but that mind is a passive target impacted or penetrated by sensory data. This deeply misguided notion puts cognitivist theory at risk. The external world is not the construction of a sensory barrage that breaks through a barrier of concepts to enter the stream of cognition. The sensory side of perceptual experience is not readily characterized in terms of conceptuality—even if sense-data are in some form conceptual—but is best described in terms of neural impulse and pattern of activation. Physical data cannot aggregate to concepts unless one is prepared to accept a like-like assimilation of mind to physical data that are conceptual all the way through. The flaw in externalism is the interpretation of objects as the source of their perception. The tree before me is not, as common sense would have it, the origin of my perception; instead, it is the outcome of a process of object-formation, the precise origins of which are inaccessible.

The relevant issue is how one perceives and identifies an object such that the raw data of sensation — nerve impulse, physical "information", bits, features — come to rest in conceptually-recognizable wholes. The greater problem is with the physical data, not the psychic apparatus, since the subjectivity of experience and the conceptual basis of all mental phenomena are given in thought and perception. This is evident in the perception of appearances, not entities, and in their continuity over time. The identity of a tree is preserved across differing perspectives, seasonal change, re-planting or pruning, even death or decay. This characteristic of objects gives the notion of essence (soul in humans), as that which makes the object unique, i.e. the basis of its particularity in spite of change. The problem of the self as a continuant, as Locke wrote, is that of the same thing "in different times and places". The preservation of identity in incessant change is also the problem of substance and process, being and becoming, the explanation of which requires that an object is not a sum of constituents or properties but a category of attributes.[7]

The category of an object transcends its momentary individuality to ensure persistence, for related instances are judged as changes within the category, not as collections of features. A tree is a categorical object that is itself a member of a category. Were an object or object-concept a bundle of physical impressions, every presentation would constitute a different object. The stability of objects is an effect so pronounced that a change one moment to the next in an object or a self is conceived as something that happens to a self or an object, not an intrinsic dynamic through which they develop.

In sum, perceptual experience is fully conceptual from the start, the material bases of which, i.e. the registration of mind-independent sensations, are outside conscious awareness. The world is not assembled into conceptual structures, concepts are not imposed on a passive

---

[7] Property-based descriptions tend to be dilute and generic, e.g. featherless biped, but not when they describe individualities. The typical determinant of identity in humans is personality, which is difficult to pin down in terms of properties. A person could undergo extreme change, even grow feathers and hop on one leg, and individuality would be preserved, but a marked change in personality leads to a change in identity. The classic example is the traumatic lobotomy of Phineas Gage in which it was said that Gage is no longer Gage.

compilation of physical data, and data from external entities are not assimilated to thought by aggregation, collation, computation or translation algorithms. Instead, objects arise out of a conceptual ground, the unconscious origins of which, though indeterminate, are antecedent to thought, image and object, more than a potential, predisposition or "bare presence". Impingements of physical data on mind delimit intrinsic process but are not ingredient in its formation.

In thought-development, a rational world is not the only possible outcome. Usually we think up a world in accord with adaptive pressures, but there are also worlds of imagination, invention, dream, animism, syncretic thinking and psychosis. The rational is closer to the external, i.e. adapted to the real, but on a continuum with its irrational surrogates. What is real and what is imaginary depends on the conformance of thought to the wider field of perception. Physical nature "says yes or no" to behavior, thus to survival; social consensus invites a conformance to the experience of others. Distortions in this process are not aberrations but approximations to the norm. The subjective aim is adaptive with varying degrees (differing qualitative states) of adequacy. More than rational thought, a deviation on the path to adaptation points to the conceptual foundations of the real. The primary difference among modes of thought is not the causal impact of the external on rationality but the adaptive parsing of developing concepts.

The mind/world boundary

Theory encircles but rarely confronts the interface of physical data with mental content. The assumption of point-to-point contact in the brain, or from sense-data to brain area, conforms to expectations of the synaptic model, as well as to an assumed causal transmission, and carries over to neuronal contacts in brain and to the hypothesis of causal interaction among mental contents. The concept of a neural system usually invokes the idea of widespread connectivity, not the more likely possibility of a field or wave front.

Though obvious, it still needs reminding that, except for activity patterns, neurons and their connectivity in the brain are essentially the same before and after a sensory impingement. At some point in maturation, connections are relatively fixed, with sense-data shaping brain activity by excitation or inhibition. Input is not a putting-in of activity in brain but an alteration of an existing activity pattern. External physical data do not enter the internal process, i.e. the brain state, but influence the pattern of activation as the state develops. This influence occurs maximally at the onset and termination of the state; mitigation at intervening segments permits the residues of internal events — feelings, knowledge, experiential memory — to come into play. The initial constraints, e.g. tectal input on pre-perceptual constructs, orient the percept-development to the outer world. This orientation is modulated by the influence of emotion, meaning and memory at intervening phases of reduced sensory input, while the final input, e.g. on sensory cortex, all but guarantees a veridical object.

Sensory influence on myriad neurons at multiple phases in the mind/brain state regulates the dynamic of the population, limiting the degrees of freedom in the derivation, which is a traveling wave from unconscious onset to conscious actuality. The main point is that the effect of sensation is to alter discharge thresholds, not replace cells on which the effect operates. The pattern of transition is an iterated conceptuality or potentiality for multiple derivations, each distributing into a hierarchy of subordinate concepts, finally to acts and objects. The process begins with a categorical prime relating to the animal inheritance and leads through images to objects. A significant input at the inception gets the process going in the right (adaptive) direction; a relative suspension of input at intermediate phases allows a personal signature to the developing state, while at the endpoint a massive influx of data ensures adaptive specificity and externalization. The object that appears to initiate the perception is, in actuality, the endpoint of an endogenous process sculpted by, but independent of, external data that, with innate tendencies and ingrained biases, channel the subjective aim.

Evolutionary Metapsychology

The concept of structure as recurrent actualization has implications for a metaphysics of the material world. If the theory holds for the process of cognition and for morphology as four-dimensional growth, it might also hold for physical entities, which would be conceived as a realization over phases or a *becoming-into-being* (Filk and Muller, 2009; Atmanspacher and Filk, 2011). There is a resemblance of microgenesis to certain themes in Buddhism and other strains in Asian philosophy, and the metaphysics of Whitehead (1978ed). If the idea has currency for theory of mind, the same process might be inferred for external reality. If one begins with the human mind as the highest known level of complexity and the basis for speculation on the external world, the process that generates mind would be a high-grade realization of fundamental process in nature. The pertinence of evolutionary pattern implies parsimony from origination to finality, raising questions as to whether features of phylo-, onto- and micro-genetic process are applicable to entities in physical nature. In a word, is the genesis of a mind/brain state a far-reaching outcome of the genesis of the universe?

Given that brain activity is material process or the realization of brain process in mind, one can ask why should the mental state represent a radically different mode of activity than the physical world of which it is part? If mind/brain process is, so far as we know, the most evolved outcome of process in nature, why should brain function differ from the evolutionary process through which it materializes? The likelihood that the mind/brain state is a realization of process in nature is supported by the origins of microgenesis in patterns of evolutionary growth, and the graded development of life forms that lead to the human brain. This assumes that a comprehension of primordial process is best attained by a consideration of its most developed manifestations.

What are the primary patterns of evolution that extend into brain process, and how is brain process interpreted in relation to evolutionary growth? A fundamental principle is that evolution is a population dynamic leading to speciation, while microgenesis is an intrapsychic process leading to specification, with ontogenesis a comparable process over the lifespan linked to phyletic and microgenetic growth trends. Among the similarities are:

1. *Whole–part transition.* This refers to the qualitative process leading from potential to actual, in fact, a multitude of such transitions in every reinstatement. This includes analogous shifts from context, field, frame or surround to item, figure, content or center, or from generality to precision, and related concepts of individuation and differentiation.

2. *From generalization to definiteness.* In evolution, the whole–part transition appears in the elimination of exuberant form. Many organisms are born so a few can survive. In brain development, millions of cells and connections are eliminated (parcellation) to achieve specificity in connectivity. The pattern of elimination of redundancy in evolution occurs in microgenesis and embryogenesis. Only those organisms (objects, ideas) adapted to external conditions survive and flourish. Redundancy in form trimmed to specialization of outcome is replicated in the mental state by a micro-temporal progression to final individuality.

3. *Constraints.* Specification is constrained from "inside" by habit, experience and the preceding state, and from "out-

side" by sensation, with the final act or object a compromise of need and adaptation. Elimination of the irrelevant mirrors elicitation of the salient. Sensory constraints on the individuation of form are equivalent to environmental pressures on evolutionary selection. Sculpting of excess gives organisms fit to survive and reproduce. In mind this corresponds to the sculpting by sensation to give outcomes adapted to the external world.

4. *Recurrence.* Evolution entails recurrence of adaptive form by reproduction of like organisms. Recurrence in microgenesis is for the mind/brain state. Progeny in reproduction are equivalent to mental states. Evolution is an inter-personal theory in which species are primary. Microgenesis is an intra-psychic theory of the origins of individuals. The life and death of organism is replicated in the arising and perishing of the mental state.

5. *Change.* Evolution is a pattern of change over generations, microgenesis a pattern of change in milliseconds. One process occurs over millennia, the other in a fraction of a second, but the same process occurs in different time scales. In microgenesis, change is the generation and replacement of states. In evolution, change is the replication of progeny. The individuation of replicates, i.e. progeny, develops in a dyad. In phylogeny and microgeny, the goal is the adaptive replication of mind or organism for another cycle of recurrence.

6. *Time.* Microgenetic theory entails the replacement of epochs, evolution the replacement of organisms. *Microgenesis — self-replication — is the foundational unit of evolution, which extrapolates the iteration of mental states to the reproduction of populations. Intra-psychic replication is transposed to extra-personal dependence. Put differently, the recurrence of individuals expands to the derivation of species. Hunger is the primary drive that sustains the individual; sexuality is the secondary drive that replaces the individual with progeny.*

7. *Selection by elimination.* In evolution, the unfit are eliminated by competition for survival and reproduction. In micro-

genesis, competition for existence is determined by the suppression of inappropriate possibilities. In evolution, less competitive organisms fail to reproduce. In microgenesis, less adaptive objects remain unborn.

8. *Neoteny* supports advance in evolution through retardation and branching at preliminary stages. This "mechanism" of creative advance (or deformity) is analogous to creativity in thought, where the dominant focus of the mental state, like that in evolution, is at pre-terminal phases.

Patterns of organic growth provide a framework for speculation on the ultimate nature of physical entities. These patterns include: becoming-into-being; phase-transition to completion; specification to actuality; the simultaneity or non-temporality of becoming; and the epochal nature of a cycle. Moreover, entities are not degraded but fail to recur. Entities generate their own time and change in becoming what they are. Every entity is a contrast of alternate possibilities, with feeling the engine of becoming and the force of actualization. The nature and significance of experience, a term that is loosely used for a variety of performances — perceptual, memorial, affective — requires a model of the mind/brain state. Microgenesis explains the different modes of cognition in relation to evolutionary theory, to what could be termed cognitive metaphysics, which presumes that the fundamental laws of mind are identical to those of physical nature.

Chapter Three

Experience

> *Whoever takes experience for his subject-matter*
> *is logically bound to land*
> *in the most secluded of idealisms.*
> —Dewey (1925, p. 43)

Introduction

There are two main strands in writings on experience, one that stops at the sensory receptors and another that engages inner events. As to the first, the notion that, as Quine or Sellars might put it, stimulation of the sensory receptors is all one has to go on, leaves experience outside cognition to consist of purely physical or sensory activity. How receptivity leads to mental activity—the bridge from external stimulation to conceptuality—is left undetermined. Moreover, if sensation—internal or external—is unfelt in paralytics, in dream and anesthesia, what is the impact of stimulation on experience that is non-conscious? Does unconscious stimulation induce behavior or thought? Some writers allege a conceptual element in receptivity. Others write of an assembly of sense-data, such that sensations are the "building blocks" of objects and thoughts. However, no part of sensation provides content for mind; rather, sensory impingement shapes the self-generation of the mind/brain state and constrains the selection or the elicitation of specific experiential content. There is no conceivable way that receptivity can deliver spontaneity or rationality, nor is there interaction between the sensory and the conceptual, which implies correspondence, not what actually occurs, namely, a parsing of irrelevancy in endogenous process as the mental state adapts to the sensory environment.

In animals and in ordinary mentality, the effect of sensation is immediate in the environmental present. With more abstract concepts not directly associated with present happenings, the effect of sensation is by way of framing thought to appeal to a social, professional or scientific community. Thought or concept-formation is enacted in verbal imagery (inner speech) instead of motility. Sensory "feedback" in language or perception does not penetrate the inner configuration but sculpts it to an adaptive outcome. The microgenetic view is that sensation is resolutely external in relation to a fully endogenous mental state. From this perspective, experience can be viewed as a system of beliefs, values, memories and other cognitive activities that are specified by sensory constraints, i.e. by the inhibition of alternative routes, but not formed, assembled or induced by sense-data.

Experience in the mind/brain state

As to this view of experience, the second strand in writings on experience, which heavily involves memory as antecedent to, and integrated with, thought, I have previously discussed some complexities of memory ordinarily omitted from studies of hypothetical components, such as short- and long-term, working etc., storage or consolidation, recognition, retrieval, the distinction of episodic and semantic, procedural and declarative, and their putative localizations in the brain. While addressing these distinctions, my interest has been on the relation of memory to feeling, thought and imagery, e.g. eidetic, creative, dream, to transitions between automatism and reflection, unconscious skill and conscious recall, subjective time and the illusory present, the role of memory in action and thought, the different modes of memorial experience and the continuity of the mental life. A detailed account of memory and recurrence in the generation, overlap and replacement of mind/brain states, and implications for the understanding of subjective time, can be found in these references, especially Brown (2010). Regardless of the final brain correlates of memory and the interaction of hypothetical components, any account will be incomplete, and probably false, without a clear description of the relation of experience to memory, thought and perception. From the standpoint of describing the mental processes that correspond to the varieties of memorial experience, the *isolation* of components or neurotransmitters and the postulation of focal brain substrates are as specious an exercise in current research as the search for a trace or engram (Lashley, 1950) was in the past.

It is a truism in philosophy that we are the sum of experience, though the exact nature of experience is left undecided. For Kant, experience was empirical knowledge based on "connected perceptions", or object-representations by way of concepts. For some, the intentional quality and/or presentness of experience is essential. Dewey (1925) asked, is experience "momentary, private and psychical" or, as he argued, does it include culture, history and empirical science? The distinction between experience as static or active, as a "storehouse" of knowledge, as immediate perception or consciousness, or as a potential for implementation—conscious or unconscious—is a feature of all cognitive systems, e.g. competence and performance, tacit and explicit, procedural and declarative, store and representation, automatic and deliberate, knowledge and action.

James (1890) defined experience as an *"experience of something foreign supposed to impress us,* whether spontaneously or in consequence of our own exertions and acts" (his italics). There is a continuous shaping by sense impressions to form a mirror of "the time- and space-arrangements outside". James goes on to write, *"I will restrict the word 'experience' to processes which influence the mind by the front-door way of simple habits and association"*, or internal relations that correspond to the external world. One can agree that the most pertinent and least arguable form of experience concerns an immediate perception, not as a passive intake but, as Bergson (1896) emphasized, an active bringing-to-bear of an individual psyche to a developing image. The "back-door" way of emotions, judgments, ideas was thought, by James, to develop secondarily. James's exposition follows Spencer, in which experience seems to refer primarily to the acquisition and origin of knowledge, including the animal inheritance, less its detailed structure or articulation in the mind/brain.

Bradley (1893) wrote that "everything is experience, and also experience is one", though oneness does not add greatly to our understanding of what experience is. For Bradley, and for other philosophers, experience is equivalent to the Absolute or ultimately real, which might be mental, material or refer to sense-data. While Bradley wrote that "the Absolute is not many, there are no independent reals", and beyond this there is nothing, others have argued for a multiplicity of reals. For James and Russell, as for Bradley, experience is given and present fact, though there is an alarming poverty of detail, which this chapter seeks to provide, as to brain or psychic function in relation to what is a present fact. Generally, experience is taken for reality and what is not experience is not real, though Sprigge (1993) raised the objection that since a conditional truth is not an actual experience it presents a challenge to those who argue that nothing actual exists except actual experience.

If actual pertains to the immediate present, does it exclude past events not revived in the present, or the likelihood or potential for future ones? Is competence in a particular skill, say the ability to play a musical instrument, or to produce a grammatical sentence, though not actual, part of experience? The practical knowledge of a grammar, or skill in any activity, consists in the physiological capacity to activate large populations of neurons in specific patterns, not to articulate the precise rules on which the skill depends.[1] The *experience* of producing or comprehending a sentence consists in the relative synaptic strengths of individual cells in differing combinations over successive phases in the mind/brain, including object- and lexical-concepts, knowledge of semantics and phonology. Does knowledge as a facet of experience include unconscious capacities that underlie consciousness, tacit events that go unnoticed, e.g. incidental learning, or unconscious beliefs, values and loyalties as implicit guides to thought, behavior and perceptual interpretation? Dream is an example of perception in the absence of sense-data, and thus the occasional appearance of non-existent or impossible images or circumstances.[2]

This is also true of the unconscious propagation of memory and the migration of feelings and ideas from one experience to another, whether shared or personal. Does reflection on an experience partake of the same knowledge base as the experience itself? If not, how does reflection differ, in relation to mind or brain, from the experience reflected on? Experience can include (the effects of) *absence*, such as deprivation, lack of love, of education or instruction, unsatisfied desires or unfulfilled expectations, false memories, hallucinations, delusions. If experience applies chiefly to immediate perception and active knowledge, i.e. thought and perception on the one hand, feeling and memory on the other, along with their spatiotemporal relations, what of events that are preserved as potential for future possibility? What of memoranda that are forever inactive or events that are forgotten but constitute the personality and influence behavior? If all events that impact the brain, as well as all conscious and unconscious events over a lifetime, are conceded to form the experience of the person, what is the value of this term in philosophical discourse?

Knowledge is a province of experience that is vetted by truth or belief, yet knowledge can be true or false, and may be outside experience except for its momentary occurrence. Knowledge that is not part of experience can be put to use in novel circumstances, as in solving a problem, creative thought, etc. Conversely, what is experiential may not accrete to the knowledge base, or if it does, may lead to a rational use of erroneous data, for example a judgment based on illusory or

---

[1] A helpful distinction of rules and mental process is a comparison of language to chess, in that a person observing two players can record the rules that govern the movement of pieces and compose a formal grammar of the game, but this would not explain the process through which an individual player decides on and makes a move.

[2] There is some evidence that nothing in experience is ever lost as in the *Lebensfilm* phenomena of near-death experience (Schilder, 1950). This was also argued from studies of hypnosis by McCullough (1965).

misconstrued experience. Belief and truth dissociate. Strong opinions are often based on false or incomplete knowledge, while truth-judgments tend to establish facts extrinsic to private experience. There is a wealth of experience behind the employment of knowledge and a wealth of knowledge behind a true statement. A true statement is a fragile raft of fact adrift on a sea of the unknown or, as Dewey put it, "the tangible rests precariously upon the untouched and ungrasped."

The self is a concentration of the experiential core of the individual, of who and what we think we are and how we perceive and define ourselves. In the conscious employment of knowledge or the adaptive value of experience, the self is the progenitor of thought, habitual or novel, a transitional segment in a passage to the purposeful or intentional from drive-based categories. A deviation in the habitual is the nucleus of the creative. Under normal conditions, adaptation forges a compromise in the final act and object. The goal of experiential memory is the replication of an event, a feeling or an idea. The importance of semantic memory — in my view, an abstraction from a category of similar episodes — is the provision of meaning and knowledge to thought and language, while episodic memory is the recurrence of experience in serial order. The order is not only for perceptual experience but applies to fantasy and dream reports. These interrelated phenomena are usually relegated to separate components in mind/brain but they are actually different segments in the progression from an instinctual past to a perceptual present.[3] Thought and emotion are woven into perception. The trace of a memory is the transmission of a configuration, not the retrieval of informational bits out of a hypothetical store. Similarly, perceptions are not deposited in a memorial bin as chunks of experience, looked up and recalled by unconscious assembly and projection; rather, sensations modify an endogenous process to mirror the external world, with objects carved out of tacit knowledge and memorial experience. The world individuates out of the experiential and knowledge base (Fig. 3.1).

This variety of experience is lost when it is equated with oneness or indivisibility, unless a uniform process can be shown to specify diverse actualities. The arguments of Bradley in this respect are not so different from those of Russell that everything reduces to brain function. Strawson (2003) continues on this line in collapsing the mental to the physical, e.g. "experience is just neurons firing." However, without greater detail on the nature of the reduction, or the relation of mind and

---

[3] In conformance with Whitehead (1933, p. 226), who wrote, "An occasion of experience is an activity, analyzable into modes of functioning which jointly constitute its process of becoming."

brain apart from identity, monism or elimination, or the nature of the brain state and its relation to different aspects of mentality, or the relation of mind to world, i.e. the role of sensation, or sense-data, in object perception, the collapse of mind to brain merely transfers the problem to another level of explanation, i.e. "future neuroscience". As James said of the soul, "Whatever you are totally ignorant of, assert that to be the explanation of everything else" (1890, p. 329).

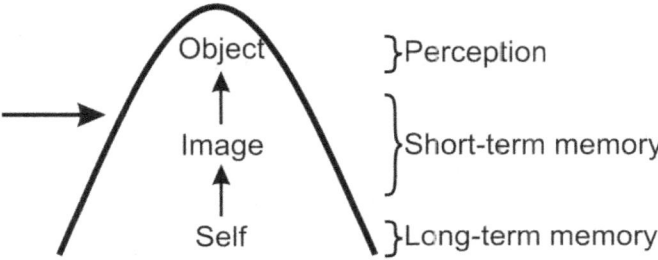

Fig. 3.1: The transition (becoming) from self to perception, mind to world, lays down the mind/brain state, with consciousness the relation of depth to surface. Visual and verbal imagery, including conceptual-feelings, e.g. desire and affect-ideas, arise at intermediate phases, but only when an external world is realized. The outer arrow represents sensation acting by way of constraints at distal segments to analyze, externalize and adapt the state to the physical world. Perception is not relayed to memory stores; rather, phases in object formation are uncovered in forgetting.

Whitehead wrote of gazing at a red patch, a *sensa* naked, stark and immediate, but except for the after-image, is this possible? In cortical blindness with object loss, there can be erythropsia or diffuse redness in a compressed visual field, but ordinarily, a colored patch is not independent of an object. Even red light covers an object or occupies a locus in the visual field. Danto (1981) put the matter directly in describing an imaginary exhibition of identical paintings of a red square, each representing a different interpretation. On his view, an interpretation is *applied to sensa*. However, memory and thought are not add-ons to objects but ingredients, i.e. *the perception of a red painting grows out of knowledge or memory*.

Experience excludes what cannot be experienced, such as redness in the color blind, or events that could be experienced but are not, such as going to Paris instead of a campsite in the Gobi desert, or the inability to walk in a person with paralysis. Experience excludes happenings outside individual mind. Should all thoughts and perceptions be counted as experiential? If I type an X in my computer and delete it, is this preserved as an experience? Ultimately, the primary axes of the mind/brain state are perception (thought, memory) and action (value, feeling), from which all mental phenomena are derived. The process of perception gives belief, imagery and thought. The process of action displays feeling and emotion. Memory refers to stages in perception; perception is externalized memory. Interest or value (Whitehead's *concern*) is the final outcome of feeling. Stout (1902, p. 225) wrote that "interest as actually felt at any moment is nothing but attention itself, considered in its hedonic aspect." If one substitutes valuation for hedonic, the argument is the same. Stumpf wrote that "attention is identical with interest." As discussed in other works, interest is a phase in valuation. The particulars of mind engage our attention but a particular is not distinct from the diachronic process through which it individuates.

Thus, a central question regarding experience is whether sensation fills the mind like an empty cabinet with "information" undergoing secondary transformation according to interest, need and learning or if an inherited repertoire of endogenous behavior that begins with drives, conditioned and fine-tuned by events, shapes experience from the beginning through revival of early phases and later revision. In a word, is the mind passive or active, receptive or predatory?

Memory and Experience

Experience conceived as knowledge is implicit or explicit memory. If experience is conceived as perception, memory is still the greater part. Merleau-Ponty (1962ed) wrote that objects are remembered into perception. The immediacy of perception conceals the richness of its memorial infrastructure, including emotion and thought. The full weight of the personal past is channeled into everyday objects, which recede back over the past to impact the next round of object-formation. The present is created out of a past that is constantly reasserted. Sensation prevents memory from achieving the veridicality of perception by ensuring that the substrates of imagery are transformed to those of objects. This leaves a floor of prior revivals as the ground of a present that extends to its anterior limit (Fig.3.2, below).

Knowledge is continuous with perception depending on the degree of revival. Knowledge that underlies thought or perception is implicit memory; thought (verbal and visual imagery) is memory that deviates from familiarity or habit. The impression that perceptions are outside the mind and that knowledge (recognition, meaning) is an attachment slices off objects as mind-independent sources of experience. However, perceptual immediacy and mind-independence are untenable given (1) the temporal lag in perception, (2) evidence for antecedent phases and (3) the duration of a "perceptual moment", with one foot in the past and the other inclined to the oncoming state. Events chunked to a world of solid objects have an internal dynamic yet to be uncovered. While the shift from one actuality to another—the sequence of conscious states of mind and world—is attributed to a causal progression, *change is within objects, not between them.* The inability to perceive transition requires stabilities as resting points. Objects appear to take flight in the world but they are nested in the mind.

The lack of consciousness for precursors—the passage from one inner space-field to another, the occlusion of the memorial in adaptation to the surround and the forward pressure from past to present—conspire in the illusion of perception as the leading edge of change. Dream foreshadows perception in the lack of past, present or future, the continuous transformation of images, and a self passive to the imagery. Dream realizes the subjectivity of thought in the objectivity of perception. Yet for all the evidence that *thought and memory are pre-terminal phases in perception,* the tendency is to begin with objects as the foundation of memory and divide experience into inner and outer.

We think of perception as the taking-in of objects and events, and postulate psychic or brain mechanisms that correspond to object-features,[4] e.g. in vision: shape, movement, color and so on. Alternatively, hallucination is incomplete striving for the external. Dream and hallucination are worlds created by thought. Objects deposit at the limit of reminiscence. One can say that perception is the skin on the flesh of imagery. The trajectory of hallucination, like that of perception, begins with implicit knowledge or personal memory and aims to objects by way of (primitive) thought. The sequence is the same in ordinary perception except that hallucination reveals this phase in the overcoming of adaptive constraints.

Time, Memory and Recurrence

We feel and believe that we perceive objects directly, unaware of their memorial underpinnings. Objects constitute the better part of a conscious state. We experience the world like a camera, the inner workings of which, to the observer, are of negligible concern compared to the picture that is recorded. Ordinarily, consciousness—of objects, acts, etc.—is accompanied by little or no reflection. When we reflect, admire, enjoy or fear an object, internal segments come into play and we are less fixated on the external situation. Apart from such occasions, perception is so emphatic it is difficult to believe the world is thought up or remembered. The separation of world from mind owes partly to a

---

[4] Bergson (1896) noted that the appearance of physical solids in the world transfers to logical solids in the mind.

lack of awareness of this infrastructure. The process leading to an object is invisible.

The idea of decay as incomplete revival entails a concept of structure as dynamic form. The standard account, consistent with our experience of ordinary objects such as trees and tables, entails an ontology of substance, external relations, object-permanence, time as a "container" and experience as a structure that persists unless degraded. However, the rust on a nail, aphids on a rose, a tumor in the brain, are changes in recurrence, with growth or involution as subjective aims. A foundational principle of microgenesis is that distributed brain systems are not static circuits but dynamic morphologies. *Every act of thought, of memory or action, is a mode of growth. Embryogenesis lays down morphology as cognition lays down thought. A qualitative whole–part shift first gives morphology as function, then function as four-dimensional morphology. A cascade of such shifts, varying in rate and locus, is a single process iterated at successive segments, not a medley of operations at different sites on different substrates and transactions.*[5]

The body changes, but time does not wear the body down. The past accumulates over innumerable iterations, pulsations, like the heartbeat. The system decays because it does not adequately re-create itself. Every state of mind and world is an arising and perishing, the systole and diastole of life as Goethe put it. Growth as memory is etched into structure by a replication that advances one step forward. What endures is reinforced by the iteration of proximal phases of tacit knowledge, with novelty at intermediate phases, adaptation at distal ones (Fig. 3.2).

As in Fig. 3.2, state-replacement and partial overlap explain the recurrence of unconscious cognition and the evanescence of conscious content. Early phases of drive, personal memory, competence and skill, tacit knowledge, implicit values, beliefs, presuppositions, individual character recur in slow transformation depending on the similarity of states that follow. Later phases relating to action, external perception, immediate and short-term memory, transient thought and feeling perish when the state actualizes to "clear the slate" for an ensuing

---

[5] I am indebted to Gary Goldberg for calling my attention to the writings of Waddington who, following Whitehead and Bohm, argued that objects are not solid things, but four-dimensional events, i.e. "possibilities of realization of certain qualities, which are realized when the objects have ingression into events" (Peterson, 2011). See related studies in morphogenesis by Goodwin (1982; 1984) and Streidter and Northcutt (1991). The process account of objects, i.e. an event-ontology, is discussed in Brown (2005) in relation to morphogenesis.

cycle. Intermediate segments — desire, thought, memory — are recoverable to a variable extent.

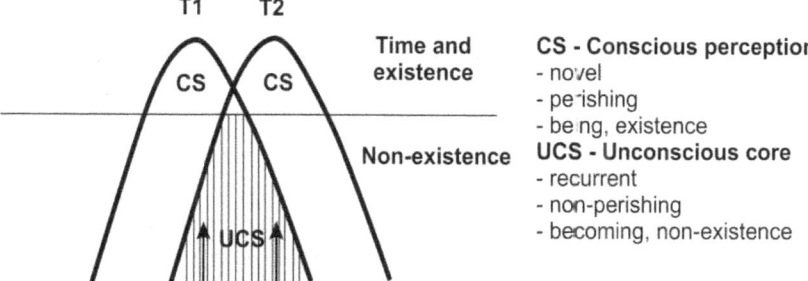

Fig. 3.2: The mind/brain state at T1 is replaced by an overlapping state at T2. The core of T1 is overlapped at T2 before T1 terminates, i.e. before the epoch exists. This explains the recurrence of early phases in T1 associated with self, character, dispositions, experiential memory, core beliefs, values and personality. Later phases perish on completion to make way for novel perceptions. The reactivation of earlier phases by the overlapping state explains the sustained attributes of personhood. Early phases are ingredient across states; later ones malleable in the shaping of endogenous process by sensation (Brown, 2010a).

Time, Experience and Life

Each day brings a faint nostalgia, a reminiscence pleasant or unpleasant, a reminder of time's passage, the premonition of Marvell's "winged chariot hurrying near", the waste, the loss, the fruits of time "well spent", the promise of futurity, the shudder at certain demise. Physical time drones on unabated as personal time flits away. We are not conscious of the passage of time but *reminded* of it. Time-consciousness is an outcome of passage in the mind/brain, but passage is not its object. In spite of the felt acceleration in aging, the sense of present duration is relatively constant. The feeling of life passing, the "running out of time", the merciless flow, the sense, as Yeats put it, that "time drops in decay like a candle burnt out", are opposed to the momentary time of overlapping nows. The bubble of the present does not move; it is exchanged. Subjective time goes nowhere. The feeling of motion and the surge to a future occur in the simultaneity of an epochal whole.

Physical time is endless, or at least a continuation over billions of years from the big bang to the projected death of the universe, a time-span that might, *sub specie aeternitatis*, be nested within a multiverse. Within this nested series, all universes, and all entities, have their time, macro to micro, from the human scale to the briefest of epochs, e.g. a chronon (Whitrow, 1972). In contrast, the subjective present arises in a fraction of a second and perishes. The death and rebirth of mental states and the disparity between onset and offset extrapolate to a longitudinal series from past to present. Simultaneity is displaced to serial order. The shift from before/after to past/present corresponds to the shift from potential/actual to cause/effect.

Anterior and posterior boundaries change but the present does not move. For physical time, the before/after is the motion of Aristotle. There is also change from earlier to later in the mind/brain state. A memory felt as past arises in an epoch in which past and present are simultaneous, like a melody heard all at once but in serial order. The feeling of pastness relates to incomplete revival; near-complete revival gives eidetic or hallucinatory imagery. The non-temporality of the state is paradoxical; succession assumes temporal order on perishing. The present makes possible an ordered sequence of events and a perception of events in change. The paradox is that change in the physical world, as in the actualization of the brain state, delivers the subjectivity through which time-experience occurs. Implicit revival gives the posterior boundary of the present (Fig. 3.3), but conscious revival is necessary for extended durations. The felt passing of ten or twenty years is an act of narrative creation. Memories are resurrected as milestones that punctuate the personal past with episodes. Remembrance generates the life-saga, while the events generated expand (the feeling of) present duration.[6]

The illusion of subjective time contrasts with the reality of objective time, but all mental phenomena — time, change, action, objects — are illusions that mirror the "out there" of physical nature. Adaptation and survival compel a simulation of natural process so accurate that, for practical purposes, ordinary perception might as well be judged as real, though for theory of mind an appreciation of its illusoriness is necessary.

Experience: Structure and Growth

Existence is sustained by implicit recall. We cease to exist when self, brain and body can no longer be revived. Pathological cognition is not loss of elements but withdrawal to earlier modes of thought.[7] The correlates of memory elude precise localization, partly because memory is not readily divided into components and partly because the term *memory* applies to a variety of performances that include storage and activation of experience, neural substrates of conscious recollection,

---

[6] The amnesic who lacks explicit memories has a foreshortened sense of the past, e.g. two months can feel like one week (Schilder, 1936; Richards, 1973).

[7] The kernel of truth in the "regression hypothesis" is that pathology uncovers preliminary process, not the behavior it deposits.

and the habits, skills and implicit knowledge on which consciousness depends. The common view is that consciousness of memory involves "holding" perceptions in a short-term "buffer" before they pass to an unconscious "store"; recall in problem-solving is termed "working memory". Memory in terms of brain function relates to the synaptic change that accompanies "consolidation" or "storage".

Subtle changes in neuronal architecture, e.g. in the configural wave that passes over phases in the mind/brain state—synapse, membrane, neurotransmitter—occur in every thought, reminiscence and perception. All aspects of memory entail a change in the recruitment, distribution and pattern of neuronal connectivity over widespread regions. Thought and memory alter receptivity to current and prior experience, as well as to other perceptions; expectancy alters consciousness, value and the impact of future events, as well as the revival of past ones. The implicit selection of thoughts and memories, the strategies and different domains of experience activated in recall, are conditioned by the immediate situation and anticipation of the future. A person thinking of building a bookcase will reflect on prior projects that, *inter alia*, activate experience relating to carpentry, aesthetic, tools, material and costs. Experience is predictive, especially for the immediate future but for long-range planning as well. Even when past experience arouses nostalgia, it is adaptive to current needs. In this respect, the meaning, significance or interpretation of an event is no less part of experience than the event in which it is ingredient. The role of experience is particularly emphatic when it induces or constrains expectations.

Conscious recall, as well as the tacit memory in every act of cognition, entails a field effect or traveling wave of progressive specificity. The dynamic in stability is concealed by near-replication. The persistence of memory in recall and its engagement in thought or action are guided by patterns of growth. The term 'decay' supposes a static trace that degrades, not deficient revival.[8] The whole–part or ground–figure transition that extends growth into cognition specifies conscious mentality out of tacit experience.

This goes to a distinction of tacit and reflective in relation to material structure, the former conceived as structural, the latter as physiological. The dynamic of recall from a memory-store is supplemented by the idea of perception passing to physiology in short-term

---

[8] Short-term memory is more vulnerable because revival must approximate perception, which recedes top-down, the inverse of the original.

memory and consolidation in long-term memory.[9] On the microgenetic account, a perceptual object grows out of "long-term" memory and passes through "short-term" memory on the way to *becoming* a perception. Stages in memory are uncovered in the reverse of the original perception (Fig. 3.3). The deviation of memory from the original event, the commingling of memory with thought in dream and perception, the elaboration, propagation, assimilation, categorization, generalization, abstraction and symbolic transformation of memory, all point to pre-object phases, and all involve growth in experience.

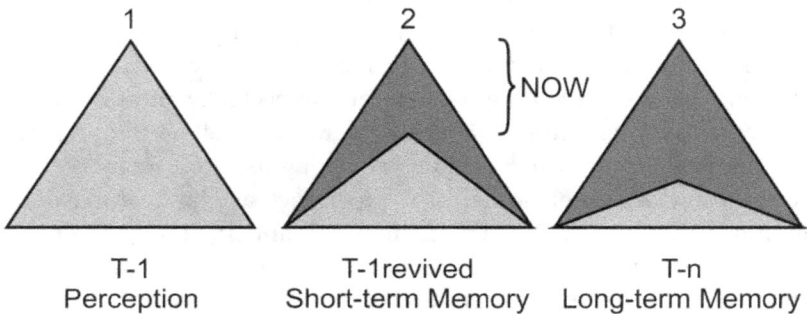

Fig. 3.3: The mind/brain state. (1) depicts the state at T-1. At (2), T-1 is incompletely revived in a subsequent state. The disparity between the perceptual "surface" and the memorial "floor" is transposed to the virtual present or now. At (2), the proximity of the revival of T-1 to the endpoint of perception determines the duration of the present.[10] The more the memory exhibits physical properties of the T-1 perception, the more characteristic it will be of short-term memory. At (3), T-1 is revived to a shallow level. The memorial content consists of the gist or meaning of the original perception, characteristic of long-term memory. An important point is that the mind/brain state develops *from* long- *to* short-term memory *to* perception, so that decay (incomplete revival) uncovers phases in the original perception. This reversal of the conventional model of memory allows a coherent picture of the relation of perception to memory and thought (see text).

---

[9] The analytic focus of science and the direction of human thought have the effect of fracturing complex functions such as memory into manageable components to simplify experimentation and reduce variables. The appeal of this heuristic is deceptive in that there is no procedure to re-unify that which has been divided. If the isolation or reduction of a sub-function is an artifact of a research strategy, so also are the results of the analysis. There is no algorithm that takes constituents, once separated, back to the higher level. In contrast, microgenesis begins with wholes and extracts derivations.

[10] The elasticity of the present owes to the degree of revival of prior states. Alteration of duration occurs in meditation and pathology. Conceivably, in memory loss, the duration of the present could be altered by a deficiency in revival.

In sum, structure is embodiment. Structure does not output memory (thought, perception, etc.); instead, memory elaborates structure. Recall is incomplete perception. Each act of thought, each mental state — conscious or unconscious — is a concomitant of growth. Memory is a mark of growth in the mind/brain that is modified and sustained by recurrence. Experience is constantly changing, in remembrance and forgetting, in scope, use, meaning, import and accessibility over the lifespan. An entity persists so long as it can be reinstated — structure, function, cell, memory, thought. Change can take the form of relative loss, gain or transformation. In each volley of recurrence, the entity is more or less reinstated. The process of cognition is itself a mode of growth with memory its residue. The revival of memory is the form taken by a momentary (which is all there is) experience. The "trace", its "consolidation" and "retrieval" from a "store", e.g. a trace that endures or decays, are the dogma of a failed theory that should be discarded.

To ask the nature or structure of experience is like asking, what is an orchestra? Is it the instruments and musicians before and/or during the music? The question is not trivial and the answer is not obvious. It goes to the distinction of natural wholes, incidental aggregates and the nature of substance. Is an orchestra a thing, a form or an activity? We speak of an orchestra as having a good or bad performance, or being a good or bad orchestra, so the orchestra changes according to the abilities of its members, just as a brain changes according to its use, form, capability and activity. What about amateur or unskilled musicians? At what point does size matter, e.g. a quartet, chamber group or large assembly? What of chaotic music or an atrocious performance? Would a group of monkeys that descended on the stage to pluck and pound on instruments be considered an orchestra? The problem is more complex when we consider that music, e.g. the sounds of an orchestra, is acoustic noise to which the mind provides pattern (Kivy, 1993). Orchestral music and the concept of an orchestra are mental objects.

If a brain state is conceived as a recurrence of the prior state plus, *inter alia*, the novelty of thought and the effects of sensation, and if the brain is the sum of its activity, as a man (Sartre) is said to be the sum of his acts, the ordinary notion of structure as fixed hardware is replaced by the growth of form as iterated pattern. The relevant structure is the field, not cells in the field. The relevant growth is the wave-front of the field, not synaptic elements. Neurons are subordinate to the configurations of which they are part. Recurrence is micro-temporal history. Structure is replacement; replacement is replication, and all replication follows the path and pattern of growth.

A Speculation on Experience and the Mind/Brain State

Microgenesis leads to an idealist philosophy that incorporates the physical world as a process that constrains cognition to a model of reality. As with other forms of idealism, concepts of (neutral) monism, dual-aspect and panpsychism are to a greater or lesser extent compelled by this theoretical stance. If it is the case that microgenesis entails a monist or dual-aspect theory, and if, as some writers maintain (e.g. Russell, 1914), the common ground of mind and brain lies in sense-data, which are neither mental nor physical, and if, further, sense-data are taken to constitute the ground of experience (incorrectly in my view), it is not unreasonable to ask if dual-aspect implies dual experience, namely, the mind's experience of itself and the world, and the brain's experience as well. Though experience for many implies subjective experience, or experience within a mental state, the relation of experience to brain structure in concepts of consolidation, storage and decay implies a physiological aspect as well. Thus the question, is there experience in psyche and experience in brain? We know what it is like for the mind to experience the world, but we cannot imagine what the brain's experience would be like. This seems to reinforce the duality of dual-aspect even if brain and mental activity refer to a common process.

One way to conceive the dual-aspect is that the mental state is the brain's experience of the world; more precisely, the mental state is the experiential content of a brain in a certain pattern of activation, i.e. a mind/brain state. The experience of finding or producing a word, and the meaning and intent behind it, are the brain's experience in a state of word-production. Similarly, the mind's perception of the world is the way the world is experienced by the brain state. *There is no separate brain and psychic experience; the psyche is the experiencing brain.*

If psychic experience is the experiential content of the brain state, is the brain state the experiential content of the mental state? While mental experience is not of neurons, chemical agents and connections, a goodly part of psyche is unconscious, the feeling of which, i.e. the feeling of activity beneath surface awareness, is the feeling of physiological brain activity. So, too, are fatigue and compulsion. The fact that conscious events in the world are felt and judged by most people to be physical and independent of mind is a way of saying that such events are not psychic or that they are physical and of the same order as brain events. Moreover, the brain itself is a perceptual object, with the observer believing that brain and object are part of the physical world. This might imply that the sense of an inaccessible unconscious is an experience of early phases in the brain state, not inaccessible phases in mind, while the external objects of consciousness, though still contents in the mind/brain state, are experienced as part of a material world. Psychic events that are purely internal — thought, dream — refer to intermediate phases. For most, the meaning of a dream takes it firmly into the mental or subjective camp, though some have argued that dreams are random neural firings. Of course, this could be said of other mental events, but dreams are closer than thoughts to an experience of the brain. Some internal states are close to physical feelings, e.g. drive, anxiety, while states such as hesitation or indecision that have a physical feel reflect the tension of incomplete specification.

Searle claimed the unconscious is pure physiology. Certainly, there is resistance to the idea that early and late phases of the mind/brain state — the unconscious and the external world — are psychic in nature. The notion of unconscious ideas is controversial, as is psychoanalysis which feeds upon them, and all would agree that we only know the unconscious when it becomes conscious, so that, with the exception of clinical data, which reveal pre-processing or unconscious phases, there is no direct knowledge of unconscious mentality other than events in conscious mind referred to unconscious cognition. In contrast, the external world is perceived as mind-independent. Yet, the former (unconscious) is early process in the brain state, the latter (act, object) is late process. That early and late process are felt as non-psychic, i.e. that conscious mind is a veil between the outer world and the inner unconscious, reinforces the argument that psyche has some experience of physical data in the brain state. Moreover, some internal events such as *qualia*, e.g. after-images,[11] have a physical *and* psychic "feel".

If the brain state has psychic experience, what is this experience other than mentality? The sense of a physiological unconscious, of *qualia*, and the perception of an external world are psychic manifestations of a material brain and, at the same time, manifestations of physical experience in the mental state. The idea of a single state with dual-aspects can be modified to the idea of dual-experience in a single state, such that the brain-aspect of the state is experienced by the mental-aspect, and the mental-aspect is the experiential content of the brain-aspect.

A description of animal mind is ordinarily satisfied with an identity of mind and brain since there is less distinction of brain and mind or behavior, even in the higher primates. Behaviors that imply mentality, such as instinct and the drives, aggression, fear, mate selection,

---

[11] After-images show changes in size conforming to optical geometry, not perceptual constancies (Emmert's law). This has been studied in cases of palinopsia, thought to be pathological after-images.

maternal care, can be attributed to instinctual manifestations and do not necessitate psychological interpretation. In humans, such behaviors can also be attributed, at least in part, to instinct and the animal inheritance. The problem of psyche arises primarily from the consciousness of self, agency and inner states. Subjectivity is experienced with delay in the direct "read-out" of intermediate and traversed segments that create a sense of individuality and personality not clearly reducible to a material brain. The separate and seemingly irreconcilable vocabularies for mind and brain reflect the description of mentality in terms of isolable substances (contents), and brain in terms of local process. However, if mental events are described in terms of process, not content, and brain events are described in terms of configuration and pattern, not neurons and circuits, the artificial divide might finally be resolved.

Chapter Four

What is Consciousness?

> *No problem can be solved from the same level of consciousness that created it.*
> — Albert Einstein

This chapter summarizes the main features of the microgenetic account of consciousness, of the transition from self to image, act and object, the epochal nature of this transition and its relation to introspection, imagination and agency. In brief, the micro-temporal transition from archaic to recent formations, i.e. distributed systems, in the phyletic history of forebrain constitutes the absolute mental state, with consciousness the relation of self to image and/or object. The discussion touches on the overlap of states, the continuity of the core over successive states, and subjective time experience.

Introduction

As I write this chapter, I am looking out at the field beyond my garden, the many shades of green in the brush, the boxwoods, the ivy, the fig and the mulberry, a dusky sky that gestures oncoming showers, sparrows darting in and out of nests, sounds that encircle me, the steady gurgle of a rivulet, the hum of crickets, frogs croaking in the distance. The object field is a whole that includes the objects around me, the "empty" space between them and the portion of space they seem to occupy, including my body and its movements, all of which are in constant change and felt to be independent of my seeing them. I am conscious of the world and those objects, as well as the thoughts, images and feelings to which consciousness is directed. Consciousness is always about something. Even with my eyes closed and in a state of passivity, I am conscious of inner or bodily feelings, a direction to thought and the inner or outer world. Unless I am in dreamless sleep, there is always some engagement with the world of objects or the imagination.

For many, consciousness is an internal activity attached to things and extrinsic to the things to which it is attached: objects, feelings, thoughts. This *consciousness-of* something is not apprehended as part of what it takes in. Objects pass and change while consciousness is little altered. We see a red object but do not have a red consciousness. Instead, we speak of a shift in attention to one thing or another, and postulate attention as a bridge from a relatively stable consciousness to changing interests and goals. This is not precisely true in pathological cases for consciousness degrades with destruction of the perceptual cortices when objects are lost, as well as in states of sensory deprivation or snow blindness. Pathology offers a clue that consciousness needs an image and/or an object.

Attention: Inner and Outer

The problem is compounded by the stark transition from mind to world. A good part of the difficulty in thinking about consciousness comes from the assumption that the mental state is exclusively interior and does not include the world, or that once an act, object or utterance externalizes, it is no longer part of the mind. We can close our eyes or lie quietly in the dark and the world disappears though consciousness persists, the implication being that consciousness does not require an external object. Yet even with the eyes closed, the external world is implicitly there, sustained by verbal and visual images close to reality.[1] When imagery is driven by fully internal events, the contact with reality is lost and the individual lapses into sleep or dream (below; Fig. 4.1). On the view that consciousness is distinct from the outer world, the connecting link is attention. However, once we depart from this standard account and realize that the outer world of perception is ingredient in a state of waking consciousness, an entirely different picture of mind emerges.

A state of consciousness is commonly parsed to a *self* that is conscious, to mental or perceptual *contents* that are its aim and to the *attention* directed to the contents. Yet the analysis of consciousness into components, which are themselves divisible—self, will, intention, desire, aim, attention—presumes that a re-combination of components gives a process that feels unitary, spontaneous and continuous. This way of thinking tends to isolate consciousness from its source in the self and its goal in the world and makes the understanding of consciousness more difficult. There is no consciousness without a self that

---

[1] Imagery refers to the variety of visual, auditory, verbal and other pre-objects, some vague like memory images, others clear like eidetic images, some fully intrapersonal, others extrapersonal such as hallucinations and dreams, but all sharing the quality of not being apprehended as independent of the self. The division is not sharp, and the contention of this chapter is that the shared "mechanisms" and continuity from image to object can support an internalist account of the perceptual process.

is conscious, and an inner and/or outer world to be conscious of. Attention merely describes this relation. An account of the relation of self to idea and object is an account of a state of consciousness.

The inclusion of the external in consciousness entails that inner segments to a varying extent inhabit external objects, even before they are consciously perceived. That is, experience and memory, concept and feeling, are not applied to an object *after* it is perceived but accompany the object at successive phases in its micro-temporal development. The postulation of a gap from inner to outer, in fact, the idea that the world is independent of mind, ruptures the continuity of the phase-transition as the final phase objectifies (Fig. 4.1). A tree is not a "naked" entity out there in a mind-independent world on which consciousness happens to alight but is an *endpoint* in percept-formation.

Consciousness of mental or external contents, including its incomplete realization in dream, can only be understood if the self is the foundation of a conscious state, and objects or images are actualizations of the distal pole of the state or its subjective or intentional aim. Whatever content actualizes is part of the attention it receives. The composer Janáček asked, why do I look at this daisy in a field of daisies? If I am looking at a field of daisies, attention and object-field are diffuse. If I am looking at one daisy, attention and field are focal. If I think of a daisy, attention is fixed on an image. One could as well say the object- or image-field is realized together with attention as part of the same state.

One aspect of attention that tends to reinforce the mistaken idea that attention is an output or activity of consciousness is the feeling that I can direct my attention to any thought or object, or evoke an image and attend to it, or suspend attention in an objectless state of meditation. Voluntary feeling can also accompany thought images, so it is not exclusively bound to action. Volitional feeling arises in the act- and image-development, just as passivity arises in percept-formation (Brown, 1996). Thought-images feel volitional, while dream, hallucination, hypnagogic and other images come unbidden. This implies that the feeling of volition is bound up with the process of image-formation, not attached to the image content. Generally, phenomena that are part of the micro-development of *acts* give an active quality, such that action and the verbal imagery (inner speech) of thought feel voluntary, while the passivity that arises with most forms of imagery gives a feeling of receptiveness, as the image-substrate shifts to an object-perception, when the passivity becomes detachment. The voluntary feeling of some images and the feeling of agency in directed attention have been addressed in prior writings, but the argument relates more to free will than to consciousness and would take us far afield the present discussion. Agency is related to intentionality with respect to the self as source or the object as goal. Images and objects seem to be targets of voluntary attention, but they also seem to evoke the attention they receive. A voluntary aim can be focused on some category or particular in a mental state or a series of states, and ordinary language distinguishes objects and attention, but a careful analysis of the experience will fail to separate the object of attention from attention to the object. When attention is directed to an object, the object, as a focus of attention, can be said to include the attention it receives, without which it goes unnoticed, even if we assume the object persists when it is no longer in the consciousness of the observer.

What actualizes at a given moment is determined by inner and outer constraints on the developing state. Inner constraints are dispositions, habits and the effects of the immediately preceding state. Outer constraints are sensory inputs at successive, but mostly distal, phases in object-formation. What actualizes is a result of these constraints, which contribute valence to an object in the form of interest. Interest is a relation of self to actuality and a mode of valuation; focal or diffuse, inner or outer, approach or avoidance. We attend to what is of interest, and what is of interest, whether positive or negative, is a sign of its value.[2] Value develops from implicit bias to conscious focus, from drive to worth. A conscious relation of the process of actualization to the content that actualizes is intentionality. Generally, to distinguish the consciousness of content from the content of consciousness, i.e. not a specific content but the mere having of a content, separates the distal from the proximal segments of the mental state. This common-sense approach is not innocuous, for it isolates a segment in the mental state from its continuance into an ensuing segment. This is like clipping a fountain in mid-stream and using attention as a connection across the divided segments. Since we attend to, or are conscious of, an object of interest, and since interest is a sign of valuation, attention can be interpreted as a behavioral marker of the valuation that accompanies act, image and object formation.

Consciousness does not, as some would have it, illuminate an object as a kind of searchlight. The image or object develops with consciousness in an adaptation to reality over successive phases. The progression from self to other—from subjective to objective—requires sensory input at the distal phase. The image "becomes real" as it externalizes. The

---

[2] The evolution of feeling to drive and desire, the creation of value from desire and its importation to objects, first as their existence or realness, then as interest and finally as worth, are discussed in Brown (2005) and taken up in detail in Brown (2012).

idea that mind extends into the world can be appreciated if we start with animism as preliminary to rational thought, not just historically but as a phase within the mental state, such that mind-external precedes object-detachment and the sequestration of mind and feeling in the individual.

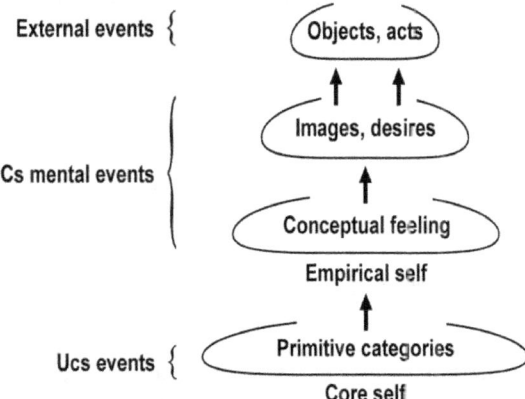

Fig. 4.1: In the mature *human* brain, the mind/brain state arises in archaic formations bound up with the unconscious or core self, instinctual drive and drive-categories (drive-representations), passing to the conscious self and conceptual-feeling, finally to acts and objects in the world. Drive categories enlist implicit values, beliefs and dispositions. Visual and verbal (inner speech) imagery is introspective content in the context of a full object-development, not a terminal addition but an accentuation of penultimate phases. Concepts in the conscious self become explicit values and intentional feelings. The trajectory from depth to surface is continuous. The final phase of objects is the distal-most phase of mind. The transition is from archaic to recent in distributed brain systems, from self to object, from mind to world, from memory-like to perception-like events, from potential to actual, from past to present and from dream and fantasy to an adaptive resolution. Through a cascade of whole–part or context–item shifts, guided by sensory constraints at successive points, a fully subjective transition objectifies a perceptual world. The entire sequence—the absolute mental state—is an indivisible epoch that perishes on completion and is revived, in overlapping waves, in a fraction of a second.

Consciousness as a Process

Let us set aside conventional thinking for a moment and allow that every mental state involves a continuous and unidirectional transition (becoming) from an unconscious onset to a conscious termination (being), passing from private space to the perceptual field in overlapping waves (Brown, 2010a). The state arises in relation to instinctual drive and categorical primitives; e.g. hunger as a drive and edible objects as an implicit category. In childhood, this phase, still unconscious, and to remain largely unconscious in the course of maturation (Schactel, 1947), is infiltrated by (assimilates) values and beliefs. A drive at the onset of the state becomes a desire further on, for example sexual drive becoming love, hunger becoming greed. The complex of concept and feeling (conceptual-feeling), or the affective-tonality of the pre-object and the categorical frame of the feeling, externalizes as an object of some value (worth). The drive *category* at the onset of the mental state becomes a *concept* in relation to a conscious self, which leads to an external *object* at the outer or objective portion of the state (Fig. 4.2).

Fig. 4.2: The drive *category* at the onset of the mental state leads through the self to the *concept* (conceptual-feeling), which partitions to *objects*. Feeling accompanies the forming object in a passage (becoming) from instinctual drive to desire to worth, conveying subjective value into the final object.

Consciousness develops in the resurgence of perception as a relation of self at one pole to world at the other. The passage from self to world incorporates implicit (unconscious) drives, values and beliefs, as well as the final, explicit content. An unconscious self in relation to drive and archaic brain formations is derived to a conscious or empirical self that is probably limbic-centered. The private space of self and image can be a terminus of the development as in dream, or an intermediate phase as in perception (Fig. 4.3). All acts and objects individuate through a qualitative transition from the core of the mental state through the conscious self to the outer world. This process recurs in every act of thought and perception.

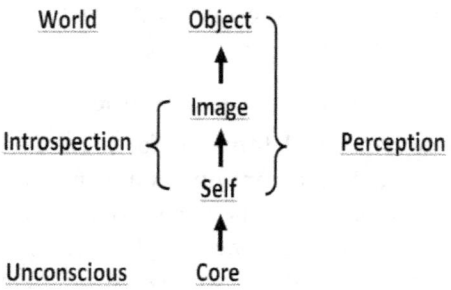

Fig. 4.3: Introspection is not a terminal addition but an intermediate phase in the transition to objects. This agrees with the evolutionary concept that novel branching occurs from earlier, plastic or less specialized formations. The figure illustrates the idea that imagery is embedded in the relation of self to world (consciousness). The relation of self to object is perception; the relation of self to image within the self-object relation is introspection or imagination. With a truncation of the final phase of object-realization, the image does not adapt to sensibility and the self is in a state of dream.

Levels in Space-Realization

Historically, the distinction of an external or physical space from an internal non-spatial field has been pivotal in the separation of inner and outer, but this assumption is in tension with the fact that mind creates a model of the outer world that develops over stages in space-realization. The trajectory from self to world in a single mind/brain state leads from a personal non-extensive space to an impersonal, extensive space. At one moment, the outer (world) is emphatic; at another, inner segments of thought and imagination come to the fore, but the entire process is necessary for consciousness. In dream, and in scenic hallucination, space is curved, foreshortened, egocentric, volumetric, viscous, almost palpable. Image and space are intra-psychic yet extra-personal (Fig. 4.4). This differs from the open, infinitely expanding three-dimensional Euclidean space of waking objects. We see in pathology a fluid shift from the real to the hallucinatory, both perception and hallucination develop over the same neural structures.

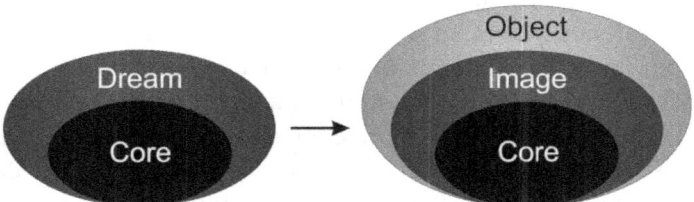

Fig.4.4: Dream is a mode of consciousness transitional to the waking conscious state. The transition from core to dream gives consciousness of dream imagery. The consciousness of dream, and the dream-self and its images, all differ from those of wakefulness, in which phases that mediate dream-imagery adapt, by the effects of sensation, to modes of perceiving closer to reality. The segment of dream imagery, though transformed, remains buried or embedded in the transition to waking consciousness. For consciousness, waking imagery, which corresponds to introspection (verbal imagery or inner speech), idea, desire, thought, requires an implicit object world.

The progression from an archaic space of the body to the proximate space of infants and the congenitally blind, i.e. the space of the arm's reach (Yakovlev, 1948), is driven by sensory input at the distal segment.

This sculpts the analysis of features, and impels the detachment of the image as an object in external space. The internal, *non-temporal* and extra-personal (yet still intra-psychic) space of the *unconscious* in dream and bodily experience develops to the *serial order* of events in external space that are *conscious* and "mind-independent" (Brown, 1983; 1988; review in 2010). The endpoint of microgenesis in dream gives an image of varying relation to reality. In waking, the neural formations that mediate dream are parsed by sensation to adapt to conditions in the world. Specifically, it is not the dream but its substrates that, through sensory modulation, are shaped to environmental contingencies. The process is wholly endogenous, with sensation sculpting the developing object to fit the immediate surround. Substrates of imagery within the mental state, even when fully realized in an object, can be reactivated in states of imagination or introspection.

Inner segments of the mental state that are the sources of feeling, idea, word and image remain largely intra-psychic, but can pass outward to the world. Some ideas and feelings rise into consciousness, then subside and are forgotten. Others find their way into action, speech or writing. Some recur; others remain incipient or propagate to novel forms. Through it all, the self is felt as the anchor of thought, the agent of action and the witness to inner and outer fields. All experience, all psychic phenomena, flow out of the self, which is the progenitor of the contents of which it is conscious.

Awareness in a young child involves a subject–object relation, and is probably similar to that in higher animals, e.g. Piaget's object- and activity-awareness. Consciousness requires a self in relation not only to objects but to feelings, images and ideas. These contents are what consciousness is about, i.e. its intentional quality. The drives — muted by implicit values/beliefs and the fractionation to subsequent phases — appear as tendencies, dispositions or presuppositions that coalesce in the core.[3] The intensity of drive, its appetitive and consummatory quality, subdued in the transition to desire and conceptual-feeling, are unconscious proclivities that implement action and thought. Every object or thought begins in instinctual drive. The intense feeling in drive, i.e. hunger, sexuality, which is sacrificed in the shift to desire and its aims, is regained in the experience of lust or extreme hunger. Tempered by the specification to ensuing phases, drive-based categories

---

[3] The core is conceived as a configural disposition of implicit (unconscious) beliefs, values, habits and early experiential memories. The core construct, presumably activated by a hypothalamic or paleo-limbic pacemaker, and influenced by the immediately prior state, is derived to the empirical or conscious self.

form the underpinnings of all subsequent phases. The entire sequence from drive to object is a complete mental state, with consciousness equivalent to the self attending to image or object space.

Microgenesis versus Assembly

Earlier phases are not abandoned to later ones; later phases are not outputs of antecedents. All phases are embedded in the termination. If someone is confused about this, for example if he believes he is thinking up the world or that objects are mental images, he is usually labeled psychotic and given medication. The upshot is that mind and world tend to be experienced in a relatively non-controversial manner, while experience in the transition of mind to world is a perturbation. A vivid experience of intermediate states can be a sign of delusion; perhaps this also applies to a theory of mind that relies on such experience as evidence. It should be emphasized, however, that a long evolutionary history of coping and survival depends on seeing a tiger as a real object, not as a mental image.

The naïve realism that takes the perceptible world for reality has to overcome the temporal lag in perception, as emphasized by Whitehead, as well as constancies, illusions, binocular disparity and a host of other difficulties. The idea that the world is reconstructed from sense-data and projected "outside" is not only implausible, but the concept of projection adds a mechanism as mysterious as those it seeks to explain. Moreover, reconstruction, recombination and projection still entail a model of the world, not a direct perception. Microgenesis is an alternative to an assembly-projection model. This theory, published close to thirty years ago (Brown, 1983; 1988), led to a model of the mental state as a becoming of earlier into later, of dream into perception through sensory constraints on the distal segments. There was (and still is) resistance to the idea that objects are the result of constraints on an endogenous process of image-development, not aggregates of sense-data. Objections to the idea that the world is an objectified dream, or an image adapted by sensation to what is "out there", have largely evaporated and microgenetic theory has achieved a certain respectability, though it's precedence is rarely cited; indeed, the transition from dream to perception is almost a given and theory now focuses on a search for physiological translation mechanisms (e.g. Llewellyn, 2011).

In sum, the essential ingredients of consciousness are a self, world and an intermediate phase of introspection or imagination embedded in the self–world relation. The subject to world relation is immediate perception or object-awareness. The self to image relation in the context of awareness is consciousness, e.g. introspection, imagination (Figs. 4.2–4). The process of world- or image-creation recurs in every mental state. Levels in space-realization lead outward from the intra- to the extra-psychic. The development is a qualitative transformation of category to concept to object. The affective-tonality that accompanies this transformation leads from drive, to desire and affect-ideas, to feeling invested in objects as value or worth. These and other features of consciousness have all been discussed in other writings.

The Unity of Consciousness

An indisputable fact about conscious experience is the feeling of wholeness or unity in spite of the claims that consciousness is constructed out of, and can be decomposed into, constituent elements. One explanation of unity is that elements are brought together by a local rhythm that binds together the modules. The idea is that an extrinsic mechanism integrates widely disparate and otherwise static elements—cells, columns, regions—in which time and internal relations have been eliminated. The process through which elements or segments are isolated—synchronic and diachronic—motivates the concept of a binding device to compensate for the loss of temporal relations and to integrate or synchronize a multiplicity of isolates. This concept extends the idea of assembly in perception to the whole of brain function.

However, the hypothesis only serves to shore up cognitivism from a basically flawed theory.[4] In microgenesis, a conscious state devolves out of an unconscious core over phases, with the entire sequence framing the state. The relation among the particulars that emanate is diachronic. Like leaves on a tree, the "connectedness" stems from a common origin of the elements and their progression to diversity through successive whole–part shifts. A central feature is that phases in the transition do not exist until the entire state is realized. The succession from before to after is a becoming into being, the phases of which come into existence when the state is realized. The becoming is the process of actualization; the being is the state that actualizes. The former is timeless, the latter temporal but perishing. The paradoxical aspect of becoming noted by Whitehead (1978ed) is expanded in Brown (2010a). As will be seen, the idea that states come into existence on completion of one cycle or phase-transition accounts for the self-identity of replicates, as well as the continuity of replicates in a world of incessant change.

Unity is an important aspect not only of consciousness but of the continuity that is the life experience. Each mental state has an insular

---

[4] A thorough discussion of the neurophysiological bases of microgenesis and arguments against cognitivism can be found in Bachmann (2000; 2006). Additional papers on physiology and anatomy are in Pachalska and Weber (2008).

character with an arising, perishing and replacement with no felt gaps within or across states. The self is felt as constant and continuous in spite of changes in thought, mood or feeling, or in growth over time. The existence of a self has been questioned from the *Anatman* theory of Buddhism[5] to the writings of David Hume, but the idea of cravings without a craver, as Danto put it, has never been fully resolved. Hume deferred an account of self-identity to future generations. James (1890) approached the problem in the idea of overlapping "pulses" of consciousness. The topic is extremely complex (Brown, 2010a) but a brief discussion is appropriate to resolve the dominant paradigm of modularity with the experiential unity of consciousness.

What would be the consequence if the mental state is an epochal whole and phases do not exist until the state terminates? The replacement of epochs is seamless, and each epoch has the character of an iterated point. Fig. 4.5 shows the overlap of states, with each state leading from self to world, from past to present, or from experiential memory to immediate perception. It will be seen that early phases (the core) of the state at T1 are replaced by an overlapping state at T2, though T1 has not yet achieved existence. The core of T1 is overlapped by T2 before T1 terminates, that is, before the epoch is established.

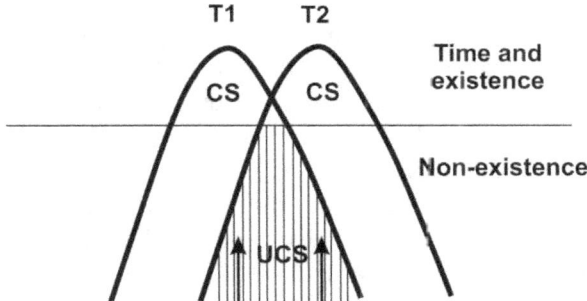

Fig. 4.5: The mind/brain state at T1 is replaced by an overlapping state at T2. The core of T1 is overlapped at T2 before T1 terminates, i.e. before the epoch exists. This explains the recurrence of early phases in T1 associated with individuality, self, character, dispositions, long-term and experiential memory, and the "persistence" of core beliefs, values and personality. Later phases perish on completion of the entire state to make way for novel perceptions. The reactivation of earlier phases by the overlapping state explains the sustained personhood behind succession. Early phases are ingredient across states; later ones are malleable to a greater extent as endogenous process is shaped by sensation (Brown, 2010a, for full discussion).

---

5   The concept of momentariness in Buddhist theory and the problem of discontinuity in successive causal pairs contrast with the Jamesian and microgenetic idea of overlap of mental states (Brown, 1999).

What is Consciousness for?

Consciousness is the relation of earlier to later in a mental state. The precise characteristics of consciousness are determined by the phase of actualization (dream, reverie, perception), the focus on inner (desire, imagery) or outer (objects, worth) and the similarity of recurrence over a series of states. Given that consciousness is in this relation, and assuming that all phenomena in the state and the relation across phases correspond with brain process, i.e. that a psychological vocabulary is equivalent to a neural one, what is the role of consciousness in mental life, and how and why did it evolve?

As to the *how*, one place to begin is with animal awareness, a relation of the subjectivity of organism to its objectified portion. Clearly, the separation of an outer world of objects is of inestimable value. A cessile organism is embedded in its *Umwelt*, but mobility requires a decision as to direction and rate of motion, exposure or confinement, safety and danger, or what is to be approached and what is to be avoided. Now the focus is on what is extrinsic to the organism and the positive or negative value it is assigned. The response to objects apprehended as extrinsic, even by tactile, chemical, olfactory or other primitive senses, fortifies the subject–object distinction. In early forms of life, one would not speak of awareness of the environment; rather, we would say that an organism is a demarcated portion of the surround. The evolutionary trend is to separation, first of the organism from the natural world, then of different objects in the world and the organism's reaction to them. The individuation of subjectivity is a preparation for further evolution to the human mind.

An early awareness of the subjective occurs in dream. A dream is an incomplete object or a memory that fails to become an event. The resemblance of some dream occurrences to those of waking perception reflects the influence of current preoccupations and habitual thinking, while the distortions reflect the lack of sensory adaptation of the dream image. The occurrence of dream in some animals is based on studies of brain activity and eye movements, but, even in humans, REMs are not strongly associated with dream imagery, so brain activity could represent the neural correlates of dreaming without a dream. Dream might occur in some animals and, to the extent that it does, it would signal, as in humans, a truncation of the mental state prior to objects. Such dream images would probably be similar to waking objects, not as in humans, distortions, fusions, symbols, or verbal narratives. Anecdotal reports of cats or monkeys reaching for non-existent objects suggest waking imagery or hallucination, but it is doubtful non-primates have images in the waking state. This is because an image in wakefulness, unlike a dream, which is a final outcome, participates in what becomes the ground of future thought and imagination. An image is a precursor to an object-perception. Visual imagery in waking animals would arouse inner representation and indicate the possibility of conscious thought.

The human newborn fused with the mother, the breast or surrogate shows gradual separation of subject and object. Initially, the mode of awareness is similar to animals. The subject–object relation is reinforced in the tendency to ever-increasing analysis. The consequence of unarrested specification of sustained whole–part shifts is the individuation of a self within the subjective field. A similar partition of the outer leads to a plenitude of objects at a point of extreme objectification. Enabled by language, further individuation carves the intrapersonal into self and image. The base of the subjective field, which is organized about drive, distills to a core construct that evolves to the intentional self of desire and partial affects. The pattern is the evolution of an unconscious core, then a conscious self in opposition to an objective, extra-personal pole that undergoes a similar individuation, for example in a multitude of objects, and object-based options, and the feeling invested in them as interest and worth. The contrast of individual and environment, or self and world, and the competing objectifications of drive and core become, with the conscious intentional self, a contrast of desire and fear, pursuit and defense.

An isolated visual image, such as may occur in primates, provides a nucleus for the visual imagination. In humans, language expands inner speech[6] to a vehicle for rational thought. Waking visual and verbal imagery signal a functional advance in the growth of the mind/brain. A withdrawal to imagery or a revival that enhances this pre-terminal phase in artistic or creative process opens the door to conceptual growth and originality. It is important to note that imagination and introspection in brain, mind and creative process arise from antecedents, not outcomes. Paul Valéry wrote that *"consciousness of possession is the secret of the inventors."* What is revived in creativity is subliminal, private, closer to the self.

Thought develops on the substrates of dream (Fig. 4.4). The active self in wakefulness retreats in dream to passivity. The intra-psychic endpoint of dream accompanies a self that is swept along by, and passive to, the image content. With *waking* imagery, the loss of automaticity heightens the sense of agency when directed outward, as in planning, deliberation and decision, or to passivity and receptiveness when directed inward in mystical insight, reverie or the creative

---

[6] See Vygotsky (1962) on the transition of egocentric speech to thought in children, Brown (2009) on inner speech.

imagination. Choice and volition (agency), passivity and receptiveness (detachment), mirror the freedom from automaticity, as novelty and uncertainty are interposed between self and world. *Thinking doesn't help thought*, Goethe wrote. Ideas come to the passive self. The truly original feels effortless.

The question of *why* is complex. To ask why consciousness evolved is to ask what it is for, what is its purpose or function. This supposes that consciousness has a function, which in turn depends on whether consciousness is independent of brain or is a particular form of brain activity or, as here maintained, is not a thing but a relation across phases. If consciousness depends on brain, is it fully reducible or identical to brain activity, is it epiphenomenal, or is there two-way causality? Does one brain state continue into another irrespective of consciousness or does consciousness intercede in brain activity? If consciousness reduces to brain, it is just one brain state after another, some conscious, others not.

More fundamentally, when we gaze at a scene such as that described in the opening of this chapter, does it matter if we are conscious of what we perceive? What is the difference between consciousness and animal awareness? The latter monitors the environment for dangers and opportunities, with shifts in attention according to need, while consciousness implies judgment and choice. Memories are evoked beyond a behavioral response. Animals make implicit choices but it is doubtful they make judgments about absent objects, or actions that are distinct from exigency. The basic question is, does consciousness instigate, accompany or follow an act, a thought or a shift in attention, or is it equivalent to acting and perceiving? I would claim that consciousness of outer perception, as opposed to awareness, entails consciousness of inner thought and memory, since to be conscious is to have a self that is conscious of something. The possibility of thinking about acting or perceiving, or the desire to have a thing or perceive an event, requires the activation of precursors in the microtemporal process underlying the mind/brain state.

Is it an error to think of consciousness as distinct from brain? If consciousness is relational, we are misled in thinking that it evolved as a direct product or consequence of brain activity, to implement brain activity or behavior, or as a veto,[7] since it is the self, not consciousness, that is the precursor or agent of change. The tendency to treat the self and consciousness either as illusory or as an entity or supernumerary

---

[7] A conscious veto suggested by the work of Libet (1999) can be interpreted in microgenetic terms as a constraint on cognition, not only at the conscious endpoint but at all antecedent phases.

phenomenon, though misguided, is understandable given their gift of exclusion from their own creations.

Awareness of the world implies a subject. Consciousness obligates a self. The continuity from subject to self, like that from drive to desire, in evolution and maturation, and the relational quality of consciousness, conform to speculations as to panpsychism. However, the evolutionary continuum is not for consciousness but for the process that eventuates in the relation that makes consciousness possible, namely, the cascade of whole–part shifts in the evolution of energy to desire that begins in archaic life forms, even embryonic growth patterns underlying the individuation of mind and world, resulting in a self, its relation to a multitude of derivations and to consciousness, which arises in the relation across segments.

This process, which begins with the transformation of energy to feeling, continues in the development of feeling to drive, the derivation of drive and its immediate objects laying the groundwork for the appearance of desire and intentional aim. This topic has been discussed at some length in other publications, but here the focus is less on the evolution of consciousness than a self that is necessarily conscious. A non-conscious self is a *non-sequitur*, as is consciousness without a self. The attributes of consciousness are modifiers of the self. With an alteration of the self in psychiatric or pathological disorders, whether a Jeckyll-Hyde or multiple personalities, the self of any state, if not degraded, will be conscious. Similarly, an alteration of consciousness in confusional states or disorientation is an alteration of the self. Consciousness is confused and disoriented, but so is the self.[8] This applies to sleepiness, mental fatigue and distraction. A somnolent consciousness is a somnolent self, a distraction of consciousness is a distractible self, and so on.

Evolution of the Self

The self evolves and matures with thought, imagery and intentional feeling. Without a self, there is no desire, no thought, no recollection, nothing of which to be conscious. There can be memory and learning in organism or in a computer, there is primitive feeling, but there is no self that learns or remembers. The distinction of procedural and declarative memory owes to the progression of animal awareness to human con-

---

[8] There are unusual cases of traumatic brain injury (Pachalska *et al.*, 2011) when an adult claims he is a child or a stranger yet the person still appears to have a normal consciousness. Presumably, the quality of consciousness would change according to the nature of the altered self, but further studies of such cases are needed.

sciousness, the latter entailing a self that is conscious of objects, thoughts, feelings, memories. For a self, to be conscious is definitional. Moreover, the evolution of the self must correspond with the evolution of desire, since there is no desire without an object and, *pace* Buddha, there is no desire without a desirer. Instinct, drive and need express the subjectivity of an organism, but only a self feels desire, and only desire is intentional. Desire is extracted from drive as the self is extracted from the core (Fig. 4.6). More precisely, the core fractionates to the conscious self and conceptual-feelings (desire, fear, love, hope, etc.). The evolutionary advantage of a self that is conscious of desire is mitigation of impulse, as drive distributes into the tributaries of conceptual-feeling. A delay in discharge transforms drive to desire and its diverse concepts and feelings, and these partitions obligate a self to know or feel them.

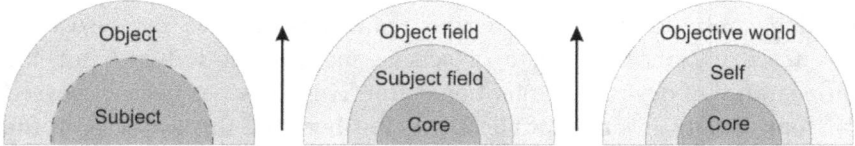

Fig. 4.6: The object individuates the subjectivity of the organism, creating a subject and an object. Then the subjective field individuates to a core and an inner field, with further individuation and separation of the objectified outer portion. The final stage is the partition of an unconscious core to a conscious self that distributes into an objective world that is apprehended as independent of the observer.

The parcellation of the subjective gives differing contents that can be treated separately, such as components of memory or language, but the direction of actualization determines the relations among the components as well as those of the self to them. Without the unidirectional pressure to actualization, there is *ad hoc* association or interaction among components irrespective of their evolutionary or maturational stage in brain and psychological growth. For example, in language production, microgenetic theory predicts a transition from the object- to the lexical-semantic concept, to phonemic and phonetic realization, not an interaction among linguistic modules.

The process of actualization that goes from self to image or object establishes the precedence of the self in every mental state, while the partition to ever more discrete particulars still leaves the self as their conscious precursor. This is essential if the self is to have a role in the implementation of ensuing events. If it were to be agreed that phases in mind (virtual or actual) flow out of the self, the question would then be whether the self is a cause, a constraint or an antecedent. The problem recalls McTaggart's (1934/1968) before-after series in non-perspectival time. Do earlier events *cause* later ones? The mind/brain state is not a box-car succession, with stages left behind *in situ* as successors appear; rather, each phase is sacrificed in the transformation to the next. This feature of the transition raises questions that might not ordinarily arise across causal pairs, such as whether a series can be non-causal, even simultaneous, with the phases ordered when the state exists. If the self does not impel, constrain or otherwise condition ensuing phases, what is the need for a self — for that matter, thought, desire or memory — all of which are derived from the self and without which the self, e.g. the *cogito*, would not occur?

How Does the Non-temporal Self Induce Change?

One explanation of the possible effect of consciousness on brain activity involves the duration of the now as a virtual span over successive mind/brain states. Prior writings (e.g. Brown, 1996) took up the question of whether the specious or phenomenal present, as a virtual arch over successive moments, might have a role in the induction of thought and action. A further possibility concerns the arousal of self and imagery antecedent to object perception, which would entail the application of constraints on a configuration midway in its trajectory. However, reflection on this topic leads to yet another possibility relating to a paradoxical feature of subjective time.

Fig. 4.5 shows the overlap of states that preserves early cognition associated with self and related phenomena. If consciousness is the relation of unconscious cognition (experiential memory, implicit beliefs, values) to content at, or close to, the endpoint of the state (objects, images), one pole, that of the core and the incipient self-concept, would be relatively stable and non-temporal, while the other pole, that of working memory, images and objects, and to some extent the empirical self of the moment, takes on existence as it perishes. Both early and late phases in the state are replaced, but the earlier ones are replaced before the state exists. Thus, early cognition "persists", while late cognition vanishes.

Could a non-temporal self influence subsequent derivations? The self gives rise to thought, intentional feeling and objects that become temporal epochs as they are replaced. The paradox is that genuine change occurs *within* the mental state, which is non-temporal, while change *across* the outcomes of successive states that no longer exist is illusory. The self influences thought and action though it is non-temporal, while the state as an epoch perishes and cannot effect the ensuing state, which has already initiated. Put differently, a state that is epochal does not cause the next, but early phases within the state overlay early phases of the preceding state, and can influence them and be influenced by them in return. This appears to be the basis of the feeling that the self initiates and controls thought and action. The sensory adaptation at the terminal phases, which may represent Libet's conscious veto, is involuntary, while the agency of the self early in the mental state accounts for the pre-activation found in voluntary motion prior to consciousness of a volition.

The revival of the state prior to its perishing accounts for stability and personal identity, while the evaporation of final contents clears the slate for new perceptions. The non-temporality of the self might also account for its elusive quality and speculations on the relation to soul and essence. The paradox is that the self is re-established in each mental state but does not actually exist, since it is laid down prior to the state as an indivisible epoch. The modification of a prior by an ensuing state, and the constraints on the ensuing state by the preceding one, offer a way out of this dilemma. In sum, agency applies to the reciprocal constraints of successive mental states, with the overlap at the onset not the termination. The subjective pole (self) effects the ensuing state by constraints on its arising. The ensuing state effects the preceding one and leads to slight modifications. One effect accounts for the novelty in renewal, the other for stability of self and character.

Chapter Five

Feeling

> ...*the individual can be cheerful and happy only if he has the courage to feel himself in the Whole.*
> —Goethe

This chapter attempts to establish on a psychological basis some foundational principles of a philosophy of mind grounded in process (microgenetic) and evolutionary theory, with a focus on the microtemporal or diachronic aspects of mental contents and the derivation and intra-psychic structure of the mind/brain states in which they are ingredient. The subjectivity of the approach is in contrast to the externalist stance of cognitivist theory, a distinction with a venerable history. For example, Bosanquet (1904) asked, "is mental growth a process of compounding units... or [as he maintained] a process of discrimination?" and cited James as preferring to begin with "the more concrete mental aspects... [and go] to elements we come to know by way of abstraction". James went on to write that the "process of 'building-up' the mind out of its 'units of composition' has the merit of expository elegance, and gives a neatly subdivided table of contents; but it often purchases these advantages at the cost of reality and truth". James insisted on a focus on entire conscious states rather than "the *post-mortem* study of their comminuted 'elements' (which is) the study of artificial abstractions, not of natural things".

For microgenesis, the process leading to a conscious endpoint is, together with the final content, part of an epochal state, the outcome of which—an act, object, word—is not a resultant of the preceding series but incorporates its earlier segments—value, meaning, belief—as part of what it is. An object includes its formative phases. The subjective has inner and outer segments. The world is the surface of the mental state. Final actualities specify pre-object phases which detach and articulate mind-external. One effect of process-thinking is a revival of the underlying continuities in the diverse aspects of cognition fractured by analysis. In process thought, wholes are potentials or categories for specification. Parts are not *in situ* in wholes but are novel derivations that serve as sub-categories for ensuing partitions.

The account of Feeling from this point of view traces conscious experience back to the physical foundations of existence, or from the facts of perception—objective data or their *appearances*—to a deeper reality. We sense Feeling in activity and passivity, or agency and receptiveness, a dynamic that underlies mentality and is accentuated when its direction is impeded, as in tension, hesitation or anxiety. If we could eliminate acts, objects or mental contents in a momentary cognition, mental activity would likely be felt as pure feeling without origin or subjective aim. The lack of direction or intentionality would suspend the feeling of before and after and result in a felt stasis of energy. The pure Feeling described in this chapter is thematic in the evolution of mind, and foundational to the derivation of instinct, drive, desire and emotion. An emotion is a complex of feeling and idea—*conceptual-feeling*—that is a motive and an object for the self. Feeling is a deeper activity, prior to emotion and idea, out of which emotion and other contents of mind develop. The isolation of actualities from antecedent possibility, the force and specificity of conative drive, the sequence that brings entities into existence, are signatures of Feeling as the engine of evolutionary advance.

Energy is the foundation of matter. At the earliest stage of inanimate entities, Feeling, as energy, is non-directional, best described in the language of physics. We can speak of Feeling when the recurrence of asymmetric energy underlies the direction and cyclical nature of organism. Feeling evolves when the recurrence of energy of an entity becomes unidirectional. Feeling, though non-relational and uniform, distributes into concepts that embody feeling as affect or emotion. Emotions such as drive or will, pain and pleasure, approach and avoidance, are vectors of Feeling that distribute into feeling, or as energy into emotion, as the essential dynamic of existence.

Introduction

Feeling is central to many philosophies, particularly those of Whitehead and Bradley to which I am greatly indebted, works that take differing positions on the nature of Feeling as the ground of existence and the relation to mental contents and those entities into which Feeling distributes. Many authors ascribe feeling to pain and pleasure with an objective component (e.g. Ward, 1920), and a subjective aspect in response to sensory presentations. Whitehead[1] conceived feeling as

---

[1] See Stenner (2008) for discussion of Whitehead and subjectivity. Microgenesis has an affinity with some concepts in process philosophy but the theory developed independently in clinical neuropsychology leading to a

an operation of passing from objectivity to subjectivity. This would agree with the present discussion if by passing from objectivity to subjectivity means the origination of Feeling as energy in material entities and its segregation in the objects (concepts) of subjective states, e.g. organisms, along with the replication of this process in every recurrence.

Bradley (1893) was closer to my account in treating Feeling as a complex unity without relations, an experience of many in one and genetically the first layer of experience (discussion in Rusu, 2013). Primal Feeling is undivided and directed energy that partitions into feeling in relation to concepts, as drive, desire and emotion (see below). Feeling does not reduce to sensation though for some authors they are identical. In the context of microgenetic theory, sensation is external to perception.[2] Those authors who relate feeling to sensation confound sensibility with perception, i.e. the physical with the endogenous, and they confuse perception and experience with emotion. In addition to the reckless employment of association terminology, the error lies in the interpretation of sensation as internal to mind and identified with stimulation, as stimulation is with feeling, and feeling is with pain and pleasure.

My position follows Bradley, in that Feeling is intrinsic, non-relational, uniform and non-decomposable. The account resembles Whitehead's idea that Feeling is a subatomic process (vibratory strings?) that, through *concrescence*[3] or *microgenesis*, actualizes the varied forms of mentality as intimations of the deeper, less-differentiated life of organism. Feeling is a quality that propels evolutionary process from its origination in inanimate nature and non-cognitive entities to its manifestation in the higher mentality, exhibiting trends in nature that transfer to the human brain as a physical entity. There is no Rubicon, or point of transition, from the inanimate to the living; rather, a continuous elaboration of Feeling into higher grades of organization and complexity.

The metaphysics of Feeling are not explicitly psychological but inevitably course through individual cognition. The contents of mind — taken as real or phenomenal — are manifestations of the Feeling that

---

novel account of time, change, process and the mind/brain state (Brown, 1988–2010).
2   The relation of sensation, as a physical constraint on perception, to perception, as an outcome of endogenous process, applies to both interoceptive and exteroceptive sensibility, e.g. pain and visual perception.
3   For Whitehead, concrescence by way of feeling (prehension) resolves the many to the one, while microgenesis postulates a progression from unity to diversity.

gives rise to them. These manifestations, such as idea and emotion, are in constant transformation. Feeling, though directional, is undifferentiated, comparable to an intrinsic energy that animates organism. The account of Feeling as distinct from energy begins and ends with an internalist perspective that traces conscious experience back to the foundations of existence, or from the actual facts of perception—objective data or their appearances—to a deeper reality. William James believed that a final understanding of psychology would be metaphysical. I would add that psychology should be the starting point of metaphysics through which, in any event, it surreptitiously passes. Metaphysics encompasses universal, indeed cosmic, wholes, but the micro-temporal history of Feeling, from top to bottom, from an individual consciousness to the immensity of space and the diversity of nature, is an account of the physical dynamic of human mentality. In a word, metaphysics is metapsychology barren of the psychological data by which philosophy should be informed.

In ordinary language, Feeling implies a relation to emotion or an affective tonality that suffuses experience and enlivens objects. An example might be the postulation of an affect-pool or libidinal stream that distributes into specific modes of cognition. Perhaps there is a relation to the Chinese *Qi*. However, a more accurate depiction is that emotions crystallize a tacit background into particular occasions of experience, some of which accompany consciousness. An emotion is a complex of feeling and idea—conceptual-feeling—that is a motive and an object for the self (see Fig.4.1).[4] Feeling is a deeper activity, prior to emotion and idea, out of which emotion and the contents of mind develop. Conceptual-feelings represent the precipitation of Feeling into affectively-charged ideas. Mental contents and events, when peeled away to expose their originating activity, reveal a convergence of matter and life that is the covert, intrinsic and impalpable quality of Feeling they embody.

Feeling is uniform but its manifestations in affect, value and emotion are protean. They assume many masks and inhabit all modes of thought and varieties of experience, differing in shape, intensity and character. Feeling is non-relational, or rather, pure relationality, and the "stuff" out of which relations develop. Objects precipitate out of the flow of Feeling. In that objects are the outcomes of a subjective aim,

---

[4] The relation of Feeling to instinctual drive, and drive to desire, is detailed in Brown (2012), which includes a lengthy discussion of similarities (few) and differences (many) with psychoanalysis. The theory is closest to that branch of psychoanalysis represented by Schilder (1951) and Rapaport (1950).

they entail directionality.⁵ All substances are forms of *purposefulness* in the relation of origination to actuality. In the human mind, the transition from core to surface, from the initiation of a mental state to its terminus, or from the onset of an epoch to its perishing and replenishment,⁶ is a relation of the immediate past of an existent streaming to an immediate present. The process leading from initiation to actuality creates a present that, in its forward momentum, prepares for an oncoming future. The sense of movement to the future in the constant replacement of mental states is such that the developing replacement gives the occurrent state a feeling of an immediate future. This is why we can never grasp the present, for as it lays down the now of the moment it is felt as the seed of the future in the overlap if its replacement, even if, indeed because, the ensuing state is not yet actual. This together with the forward progression from memory to perception, or from past to present, and the disappearance of the present in the oncoming state, give the feeling of a future constantly revealing itself as idea and presentiment.

The present is delivered out of the personal past on the way to perceptual adaptation. Possibilities envelop and individuate the final datum. Need adapts to necessity as contents are "selected". The present, or the outcome of the present state, resolves the contingency of conceptual possibility with the uncertainty of external events in a changing world. The specification of acts elaborates the choice inherent in agency. A conation to the present is the seed of purposefulness, reminiscent of Whitehead's insight (1933, p. 249) on an occasion of experience, that between an effect facing the past and a cause facing the future lies the teleology of the universe. The present surges into existence as a forward impulse to satisfaction, advancing the inheritance of an onto-phyletic past in a traversal through which objects resolve out of change to create a novel universe.

Evolution and Panpsychism

The teleology of Feeling is a speculation on the final aim of evolution. The idea is that anisotropic Feeling empowers evolutionary advance. The continuity of Feeling and distribution to affects, together with the

---

5   For Whitehead, the subjective aim is the direction to value. Here, the term is used similar to intentionality in the aboutness of the mental state, its progression to an act, object or idea. An object is an externalized image with its affective tonality, e.g. value. Value is not attached to objects or projected on them, but is specified with the object in its momentary journey from mind to world. The link is Dewey's argument that facts are irreducible values.
6   Michel Weber put this succinctly: actual entities "come into existence and then sediment into being (not vanish into nothingness)".

basis of substance or being in the epochal nature of process (becoming), are sufficient to account for evolutionary gradualism as successive stages of proto-mentality without the postulation of emergence, e.g. of consciousness. The higher levels of realization to which evolutionary process leads are not attractants to their attainment but adaptations of Feeling and its implementations in the striving of antecedents to a further level. The account is consistent, though not dispositive, with regard to intelligent design. The intrinsic activity that underlies a surge to finality, the relation to embryogenesis and growth, the recurrence of successive epochs, intrinsic relations and the conformance of microgenetic and evolutionary pattern, mark a wave of inherent purposefulness or proto-intentionality that tends to be obscured in a sea of contingency.

Energic process, displayed in the transition from whole to part, or from generality to definiteness, carves particulars out of categorical wholes. Contingency and chance are the *externalist* attributes of "adaptive strategies" that cede change to the environment, e.g. elimination of the unfit, while the *internalist* response is that unidirectionality carries a pending aim to realization. The isolation (sculpting) of an actuality from antecedent possibility, the specificity of conation or drive, the sequence that brings entities into existence, are signatures of Feeling as the engine of evolutionary advance. Energy is the foundation of matter. At the earliest stage, Feeling is non-directional and best described in the physics of elementary particles. In the gradual shift to vegetation, energy assumes directionality in primitive life forms, or perhaps one could say the life forms channel energy in the direction of becoming and/or growth. Birth, growth, death and re-birth exhibit direction within recurrence. The asymmetric manifestations of energy combined with successive recurrence underlie the direction and cyclical nature of organism and physical matter. Every entity, object and mind/brain state is an epoch of becoming into being.

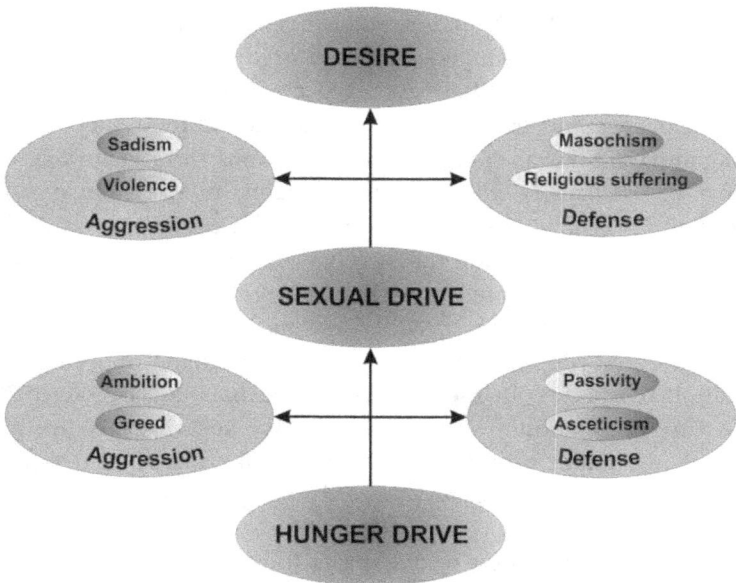

Fig. 5.1: The core drive of hunger (thirst) replicates the organism. Hunger distributes into the opposing vectors of aggression and defense (fear, flight). These vectors underlie self-preservation in feeding—in predation and in escape. Hunger is derived to sexual drive. Some affective states relating to these drives and their vectors of expression are indicated in the figure. The constellation of hunger and sexual drive gives rise to desire, which is implemented in acquisition and avoidance.

The account of evolution as a population dynamic centered on group-speciation contrasts with the microgenetic interpretation of the specification of final existents—particles, organisms—in relation to which population effects are secondary. Put differently, the societal thrust of evolution is empowered, generically, by subjective aim, potential at onset but developing in conformance with external conditions in the assimilation of drive to adaptation. Pleasure and avoidance of pain are goals for objectification or constraints on deviation in the path to satisfaction. The guiding principle is survival of the fittest, but the first priority is re-instantiation of the organism. Self-preservation is self-replication, which entails the recurrence (? causal persistence) of the organism. The motive force of recurrence is hunger,[7] which is prior to

---

[7] See Brown (2012) for discussion. The hunger/thirst drive leads to feeding and avoidance, which promote the survival of the individual, while the sexual drive, which appears later in maturation, accounts for the survival of the population. The recurrence of the individual is prior to the replication of the species. The chicken recurs each moment before an egg for the future is laid.

sexual drive (Fig. 5.1). The latter entails engagement with others in the service of the population, replacing the organism with progeny. The perpetuation of species—their persistence and change—is an outcome of the recurrence and perishing of individuals and the renewal of overlapping and continuous epochs—representations—of the world.

From this standpoint, the population dynamic in speciation is a group model of an intrinsic process of self-actualization. Species are multiples of individuals. To step outside the survival of the individual to the survival of the population preserves the pattern of specification even as it externalizes the intrinsic process of individuation. Microgenesis expands the renewal and adaptation of individuals to the renewal and adaptation of species. In the species, individuals can be sacrificed as long as the population endures. Sexual drive sustains the group, and thus has a particular prominence. But from the individual perspective, the preservation of the solitary organism is paramount with hunger and feeding primary.

The basic manifestation of Feeling is the existence of an entity. Its most elaborate articulations are the intellect and values of the human organism. The continuum from the smallest particle to the complexity of a material brain begins with *intrinsic* value in material entities. Rocks and organisms are differing patterns and complexities of atomic units. The intrinsic value of a rock, its existence, is the energic process that generates the entity, and through which it recurs. Intrinsic value in organism is analogous to that in physical entities. A rock that is replicated (replaced) over some duration is a far-distant precursor of human cognition. In the physical world, a rock is an *aggregate*, a "society"[8] or compound of atomic particles. In perception a rock is a *whole*, with the potential for realness, meaning and value (as worth), e.g. for a wall, weapon. Feeling runs through all things great and small. Some suggestions as to foundational process in nature continuous with consciousness include an uncollapsed wavepacket, virtual photons (Romijn, 2002), quantum entanglement (Shields, 2009) and microdurations (see critique in Hunt, 2001). Dombrowski (2001) has written of the "microscopic sentiency found in cells, atoms, and particles".

A particle is a basic object that science represents as self-identical over time or context-independent. Yet a proton in a stellar mass is different from one in a hydrogen atom, as a brain cell in a tissue culture differs from one in an active brain (Birch, 1990). An elementary particle can be conceived as a waveform of energy that is epochal or, if quasi-epochal, developing out of a space-time continuum such as Bohm's implicate order. The epoch is the temporal extensibility of the particle,

---

[8] In Whitehead's terminology: a "corpuscular society" (Cobb, 2008).

i.e. the minimal duration for the particle to exist. Energy condenses to a particle; duration is an epoch or category in the momentary life of a particle, over which the particle becomes what it is. *The process of Feeling that accounts for the existence of an object – particle, brain – is its intrinsic value.*

In my view, evolutionary thinking not only opens the door to but obligates some form of panpsychism.[9] The distinction of panpsychism and/or panexperientialism from emergentism (see Strawson, 2006) defends a continuum against assaults that disavow conscious experience in material entities. The skeptic will contest the argument that rocks or particles have experience, psyche, or a primitive mode of consciousness. But the postulation of a continuum merely implies precursors of higher cognition. Energy fills this need. The alternative to panpsychism is sudden mutation or a saltatory leap[10] from one mode of perception to another or a spandrel that is epiphenomenal to other adaptations, e.g. language, but even if true, this does not exclude the need to explain how consciousness arises, whether the crucial step involves genes, cells or connectivity patterns and/or complexity.[11] Does consciousness appear at a given point? Is it an emergent? If so, what is the distinction of an emergent from a resultant, the causes of which are unknown? The fatal presumption takes consciousness as an abstract entity that is the goal of the inquiry rather than describing phenomena that make consciousness possible.

Intrinsic Value and Existence

In sum, bi-valent energy interior to a particle evolves to uni-valent Feeling interior to organism. The step from physical matter to living organism transforms isotropic energy to anisotropic Feeling, with one cycle (packet) of energy constituting an epoch of existence. The intrinsic

---

[9] There are similarities with the views of William James, who advocated a form of panpsychism in which reality consists in innumerable flows of feeling interacting with each other. Human consciousness is one kind of flow of feeling typified by the high level of conceptual thinking it contains (Sprigge, 2005).

[10] The theory of punctuated equilibrium may be applicable to this problem. Though popularized by Gould, I recall this idea discussed well before by Richard Goldschmidt, my genetics professor at Berkeley, based on studies in drosophila.

[11] Size alone is not explanatory, since the brain of a gorilla is larger than that of some bushmen and dwarves (Dart, 1956; Lenneberg, 1967), while complexity has to be parsed in terms of spatiotemporal pattern of activity, not mass and intricacy of connections. Complexity is not an explanation. The Beethoven 5th played backwards is complex noise.

nature of the process is the seed of higher subjectivity. In basic entities or in brains, Feeling is non-relational, since the relata — the onset, terminus and phases of the energic wave — are not interrelated segments or polarities. We would not say, except in a trivial sense, that in a fountain the initial jet of water or an intermediate segment relates to the spray at the surface, since the former become the latter. The relations of feeling embodied in categories are over phases in the mind/brain state, and their overlapping replacements, while Feeling is the non-relational ground out of which these phases develop. Phases in the mind/brain are non-temporal (simultaneous) until the state actualizes with relations between phases (whole-part transitions) in abeyance until the sequence terminates. The question of external relations is beyond the scope of this chapter, which concerns intra-psychic or subjective phenomena.

It is a long way from a particle to a brain, yet the *pattern of process* is comparable. The brain is a complex *physical* entity, existing like a particle as a duration of constituent phases. A brain state is a hierarchic series of vibratory patterns that pulse each epoch into existence. The pattern of neuronal activity over phases — rhythmic, oscillatory, cyclical — is analogous to the vibrations of a particle (Gunter, 1999). A neuron exists as the momentary envelope of its activity pattern. We have no precise knowledge of psychic experience associated with a neuron, nor for that matter with the innumerable neurons in any region of the brain, nor the presumably quiescent neurons in a sleeping brain. A complex pattern of activity is essential for cognition, not only a *spatial* configuration but a *temporal* sequence in large neuronal populations over distributed systems in forebrain evolution.

The argument is that the *intrinsic value* (*existence*) of physical entities consists in a waveform of energy over their temporal extension, i.e. in Whiteheadian terms the minimum duration of process needed for entities to be what they are. In this bare epoch of existence, the waveform of the thing — its vibratory or oscillatory structure — is non-directional or isotropic. Gradually, internal relations expand as the nucleus of a shift from energy to Feeling; process takes on direction and becomes anisotropic (Fig. 5.2). Internal relations in the epoch expand. The duration of process that constitutes existence enlarges as intrinsic value.[12] Existence becomes *realness* as the ground of later derivations, e.g. drive, instinct, desire. With respect to value, which includes interest, the existence of an entity precedes its realness (not reality), which leads to object-worth (Brown, 2005).

---

[12] For historical precedence for the idea that existence is the initial step in value-creation, see Perry (1926).

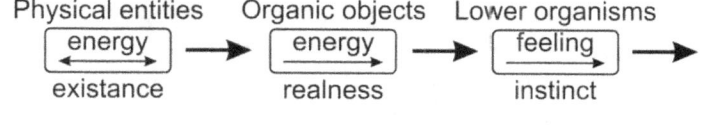

Fig. 5.2: The evolution of Feeling begins with the isotropic vibrations of a particle. The energy in these entities becomes directional, anisotropic or time-irreversible. Feeling, or value as existence, becomes value as drive and emotion. Existence as a packet of energy becomes life as a vector of Feeling. An entity consists of the temporality of its process and the spatiality of its category. Feeling as process, category as substance, are modes of *becoming and being*.

Relations

As noted, the idea of relations implies things or terms that are related, but from an internalist standpoint the relations of things and terms are outcomes of the process through which, as momentary exemplifications, they come into existence. Put differently, process lays down the terms (categories) it is presumed to relate. These pass to ensuing categories but do not constitute terms in relation. There is a temporal dimension in precedence or simultaneity, yet as mentioned, the relationality of endogenous process is non-temporal until it actualizes. While the Feeling that generates the state is non-relational, earlier and later phases in the state, which are articulated by feeling, have a relational quality in respect of their succession, but the phase-transition of the mind/brain state cannot be punctuated by a temporal locus. One can only say, on completion of the transition, that A precedes B, but A is not in the past of B, since A and B are co-temporal in the epoch. Moreover, a locus between past and future within a state requires that a series of mental states generate a present, or perspective, one that is felt in the present and one that is felt in the past. This past-present is an ascription that assigns a past-point that breaks earlier-later, with events facing the past of the point or its future.

An actuality that represents a succession, e.g. a thought, an object, is an epoch or series of epochs. Indeed, all objects are events, or clusters of epochs. The temporal order of ingredient phases becomes real when the epoch actualizes. In this process, the greater part of the mental state is devoted to recurrence, either as the underpinnings of the state or as memory. Except as an eternal object, the past *exists* only in the present.

Each mental state and each state of the world replaces—and thus in some sense, remembers—the preceding one. The transition from before to after in a given mind/brain state is comparable to physical passage in the world, while the present gives a perspective for a conscious time series, i.e. the A and B series of McTaggart (1901). A sequence of perceived states, i.e. the perception of change, is, as Bergson (1923) noted, wrongly inferred as a kind of horizontal line extracted from the replacement, whereas continuity owes to the *superimposition* of an epoch on its precursor and the *transposition of epochs* to perceived successions, i.e. to a sequence of final actualities. How the simultaneity of a subjective past, which is embedded in a present, unfolds to the serial order of events—inner and outer—is a complex problem discussed in Brown (2010a). The shift from one temporal series (before/after) to another (past/present/future) corresponds with the actualization of the state and the perceptible sequence of completed states. *In this shift, the authenticity of internal relations within a state transforms to the illusion of external relations from one state to the next.*

Feeling generates an indivisible *cycle of becoming into being*. Think of the inability to divide upstream and downstream segments of a river to relate one segment to another. A stretch of water may seem to correspond to some arbitrary point, e.g. the river bed, but the stream is a traveling wave in which one segment becomes the next. In a stream, as in microgenesis, or in the actualization of any entity, *the earlier becomes the later*. One would not say an orbiting electron in a hypothetical atom relates to, or is a relation of, one point to another. The motion of the electron creates the relation, i.e. the trajectory establishes the relation rather than the relation determining the position. An atom cannot be said to exist in the absence of a complete orbit. Similarly, an entity cannot be sliced in time but exists when it *becomes what it is over its becoming*. Feeling elaborates things to be related without relating them. This raises the question, what are those things and how are they created, and does Feeling change with changing objects and, if not, how do objects change if feeling is unchanging?

To repeat, Feeling—as opposed to feeling—is non-relational, or not relational in the ordinary sense (of external relations), in that it is continuous throughout one cycle of actualization. The relationality is constitutive, not interactive. The cycle forms an epochal or modular whole in which the succession within the state is a continuous becoming, while continuity in the succession of states is by way of their overlap. A state, epoch or actual object does not, at least not in its conscious appearance, cause the next to occur. Instead, the object is relinquished in the replacement. The epoch is simultaneous with the phase-transition that configures it, such that it exists before the succession has temporal order or exists in time. Phases in concert with the entity as a whole are condensations of Feeling, while the final epoch—emotive, linguistic, etc.—embodies, or is a category of, internal relations of the entity. Such relations, like sine waves, are purely relational without things to relate to.

For example, the desire for an object can be interpreted as a relation between the self that desires and the desired object, but the object (idea, concept), whether perceived as internal or external, is part of the same mental state as the self and the feeling of desire. As the subjective aim of the state, the object does not stand in relation to earlier phases but incorporates and actualizes the entire sequence. The object consists of all of the phases through which it develops, so that value and meaning are not attached to, or projected on, an object after it is perceived. The growth in value of an object is coincident with its replacement; the object does not remain constant while its value changes, it changes with its value in each perception. This is the import of the epochal nature of the state. Similarly, an utterance or perception may seem to consist of separate components—conceptual, semantic, phonological—and subsume a variety of processes and contents, but the components (categories), which are manifestations of the Feeling that distributes into them, are incorporated in the trajectory.

More precisely, Feeling obtains in the internal relations of objects and organisms as the process through which they objectify. Value arises in the transformation of existence to realness.[13] With an expansion of subjectivity, realness becomes attention or interest and, finally, is exemplified in drive, desire and object-worth. In the actualization of the mental state, Feeling bifurcates into subject and object, filling and imbuing the observer and, *inter alia*, infusing the object with realness.[14] Interest is value sequestered in an object. It is a thread of mentality that "connects" mind and world, i.e. an objectified portion of subjectivity in a perceptual target. In the partition of inner and outer, and in differing proportions, Feeling is allocated to self and other, with the latter a tributary of the self that does not so much attract emotion as enjoy it.

Emotion

An account of the process generating an object or mental state as directed Feeling must address the categories—internal and external—

---

[13] Realness is the quality of appearing or feeling real. Images can seem real, e.g. dream, hallucination, though not object-like in their realness.
[14] A withdrawal, or de-objectification, of the final phase of the mind/brain state leads to de-realization or loss of the feeling of realness, in objects and, especially, in other people, even for one's self (Brown, 2004).

into which Feeling is directed. How do categories arise, and what is their relation to Feeling? The categories most closely related to Feeling — the drives and their specification to the emotions — is a good place to begin. The energy that evolves to Feeling leads to instinct, drive and the aggressive and defensive vectors (Fig. 5.1). These qualities of Feeling are sediments of process, not energy in association with idea. They constitute the form taken by the becoming, i.e. the components and behaviors into which Feeling distributes. An emotion collapses Feeling to a specific affect within a category. When an emotion or thought objectifies, the category embodies the Feeling that pulses in the background as its source and incitement.

Whitehead noted an "analogy between the transference of energy from particular occasion to particular occasion in physical nature and the transference of affective tone, with its emotional energy, from one occasion to another in any human personality" (1933, p. 242). Strictly speaking, the analogy is a likeness due to the progression of physical energy into organic Feeling. Emotion is a later stage in the evolution of energy, or a higher grade in the ramification of Feeling. Any emotion objectifies personal feeling together with a concept, the varieties and complexities of which obscure the authenticity of Feeling as the primary impulse of organism.[15]

Feeling considered apart from the objects and emotions that are its derivations is not a content in consciousness but prior to its differentia, enlivening the things and events into which it distributes. The acts that for some define a person, together with the inner life that for others is the greater part of individuality, still do not adequately represent the deeper current from which, like froth on the surface, evanescent behaviors are specified. An act of cognition seems to be experienced directly, but there is an unfelt lag in its conscious realization. Moreover, the feeling in an act or idea may seem outside the contents that are felt, though Feeling, which is not itself felt, is what feels them, as modes of subjectivity to which self-realization refers.[16] Only with the intensity of early cognition is the feeling within an act inseparable from the content,

---

[15] The partition of feeling into the variety of emotions and affect-ideas is a complex topic outside the scope of this chapter. An attempt to deal with this problem, along with a critique of Freud's account and the James-Lange theory, is in Brown (2012).

[16] The widespread effort to more deeply understand the genuine or authentic self (Brown, 2005), and the variety of methods used, from psychoanalysis to meditation, points to the artifice on which a life is constructed, the intuition that mind has not been fully plumbed in thought or action and that depth is not content but Feeling, namely that human behavior cannot realize the inaccessible truth of an unconscious mentality.

though even then, especially in higher organisms, there is no doubt some unfelt antecedent lag. The "I" that loves or fears, even if overcome on occasion, is generally felt as distinct from loving and fearing; the lag begins before the category of the self — the "I" — develops.

Uniformity of Feeling decants to locality in acts and objects. To feel is to feel a relation between the actualized state and a ground that is itself largely unfelt. Experience is for or within the local content into which Feeling distributes. Even agitation gives actuality to feeling as a rupture of uniformity. Admittedly, it is unclear what evidence would suffice to turn what is obligated by a theory into a fact that supports a truth. Wordsworth alludes to the precedence of belief over fact in:

> And 'tis my faith that every flower
> Enjoys the air it breathes.

Feeling and Category

The encapsulation of feeling by emotions that express concepts in diverse exemplifications is a transformation of invisible uniformity into palpable actuality. Each mental state is modulated by constraints of the just-prior one, by ingrained patterns of process and connectivity that relate to habit and character, and by adaptation to the world or the adjustment of need to circumstance. In the human mind, hunger and sexual drive partition to desire and intentional aim. With an expansion of this trajectory, there is an opportunity for segregated accentuations within a fully-realized epoch. Drive, desire, interest, worth, all modes of value, are the dominant affects that specify ensuing phases, e.g. drive to desire, or individuate a designated focus, e.g. desire to love, fear, hope, etc. The final object also undergoes specification as attention or value shifts from whole to part, for example, attending to a person, a face, eyes, voice and so on. There are fluid shifts from category to subcategory, e.g. drive to desire, and the reverse, not as a regression but an incomplete revival.

A category is a set of actual or potential objects that are related by shared attributes, but it can be thought of as an envelope that frames an entity, an elementary particle, a mind/brain state or a segment within the state. In this latter sense, in framing the micro-temporal development of an act or entity, the category is the being or "substance" of the entity, while the Feeling within the category, the micro-temporal transition through which the entity develops, is its becoming or process. The relation of mass to energy foreshadows that of substance to process. This relation is replicated in the categorical primitives that enclose a drive, or in the conceptual-feelings that embody affectively-charged ideas or objects.

In prior writings, I assumed the process transitioned from one phase to the next, but I would now argue that phases constrain the passage of a single wave, with traversed segments isolated after the transit becomes actual. The transition cannot be construed as a causal output of one phase to the next; instead, the traversal, which is continuous, is submitted to constraints at successive phases. A mental content—inner or outer, category and Feeling—is a phenomenal outcome that, for the moment of its appearance, is an endpoint in the partition. *Thus, the response to the question "what is the relation of Feeling to category?" is that the category embraces formative phases that constitute the being of an entity, while the process of Feeling embodied by the epoch is its becoming* (see Wallack, 1980).

In the evolution of mind, becoming undergoes an expansion of the inward subjectivity of organic life. Successive phases in actualization form sub-categories within the whole of mind/brain. The category of drive corresponds to a population of neurons and connections configured by synaptic strengths that sequester Feeling at its inception. Innate patterns of activity draw in or concentrate Feeling rather than Feeling conjuring up behavior. Similarly, in the derivation of drive to desire, the concept (category) accompanies an influx of Feeling, such that one can speak of a strong or weak desire. But a concept is not a container of Feeling. A drive can swell with feeling and discharge, or transition with a quota of Feeling to desire. A portion of residual Feeling remains in drive as the engine of desire; another goes with the conceptual partition to desire. Indeed, without Feeling, desire would be a non-directional idea, like a dictionary entry. In sum, an affect-laden concept actualizes a subjective aim within a mind/brain state. Images and objects are derivations of earlier configurations that receive Feeling from a source-drive, as objects receive implicit beliefs and tacit knowledge from source-categories.

In human mind, the more primitive category of drive and its concentration of Feeling partition to the sub-categories of self, desire and the objects or images of desire. The quality of Feeling is uniform within the state, though at one moment it is emphatic in unconscious drive, at another, it is pronounced in the conscious self and desire, and then, subdued, is carried through to its objects. These transitions, markers of value in the dynamic of Feeling, dominate a mental state for a moment and pass like eddies in a stream.

The distinguishing features of a concept or category are the lack of precise boundaries and the virtual or categorical nature of content. Every member of a category is still a category of subordinates, which in turn are categories for further partition. A category can be innate such as instinctual drive, or acquired and spontaneous with its own affective charge, such as "things to take to a picnic". A category, which gives specificity to the feeling it incorporates, "splits" into subordinate categories in a forward development, or "descends" to a source that is essentially bottomless. The contents of mind — ideas, emotions — vary with the categories and their feeling-tone. Feeling gives force and valence to ideas, while ideas embody emotions and give variety to the mental life. Weak and strong feeling, such as affection and passion, reflect an influx of Feeling, which must be reconciled with the fact that Feeling assumes specificity by virtue of the accompanying idea, i.e. a difference of degree becomes a difference of kind.[17]

On this view, Feeling accompanies the category *ab origo*, distributing into ideas as local affects or emotions and constituting the affective quality of concepts. A core category is partitioned to more refined concepts and affects. A specific affect or emotion along with its concept, e.g. envy, pride, humiliation, etc., may seem a distinct affect-idea, but it is a tributary of Feeling that invests every aspect of the mind/brain state in the fractionation of drive to partial affects and ideas (conceptual-feelings).

An organism is a unity and a multiplicity. For Whitehead, a society of parts is prehended in the concrescence of a novel individuality, a kind of fusion by way of feeling of the many into the one. In the generation of a mind/brain state, myriad units along with their own pattern of Feeling assimilate to larger organs, like the strings of one violin in an orchestral piece of music. The "gathering-up" of elements into oscillatory patterns in the before/after of the mental state goes from whole to part, or from unity to diversity. This transition underlies a shift from the purposefulness of drive to the manifold of desire, and its partition to acts, images and objects. Mind reaches outward to create and fulfill its own diversity as Feeling binds disparate elements. Multiplicity at the outcome of the state unpacks potential at the base.

It is likely that the vibratory foundation of minute elements, e.g. kinetic energy in cellular units, combines and transforms to virtual oscillators (Fig. 5.3). The progression from a foundational or fundamental frequency to a series of harmonics at levels in speech production (Brown, 1988) provides a model for prehension as a reverse of this sequence, i.e. from the smallest units of organism, or neurons in brain, to a rhythmic oscillator parsed to successive frequencies. In a

---

[17] Alternative accounts in which feeling attaches to, or is attracted by, an idea, or that feelings are specified prior to attachment, or that ideas "find" and combine with appropriate feelings, i.e. that Feeling is a composite of feelings, are unsustainable (see Brown, 2012, for discussion and critique).

word, it is vibrations all the way down or, as Heraclitus put it, the road up and the road down are the same.

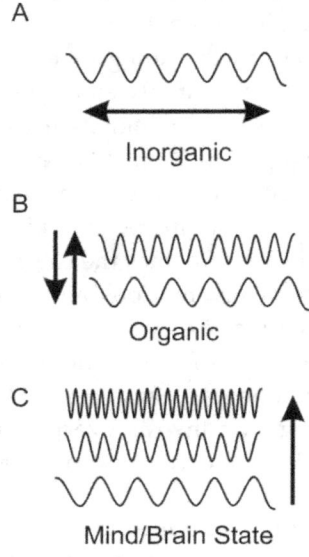

Fig. 5.3: The transition in evolution, and in the momentary genesis of a cognition, from isotropic energy to directional Feeling in organism to intentional process in the mental state. The transition can be depicted as a series of oscillatory or harmonic levels.

Becoming

A theory of Feeling is a psychology of becoming. The postulation of becoming in opposition to an ontology of being is a distinction that goes back to Heraclitus and Parmenides. The nature of thought, which is that of increasing analysis, favors substance theory in the fractionation of wholes into parts, and the solidification into observables of the invisibility of transition. Substance is palpable, while the process basis of substance has no boundaries or resting points. A quantitative psychology in which objects are stabilities avoids the change in objects, or how a changing object is recognized as the same. Walking man and sitting man are the same man, as is the man who grows, ages, adds a beard, is healthy, sick, or differs one moment to the next and over the lifespan. An object seems to remain the same in spite of incessant change. This way to think about objects appeals to common sense; if every change no matter how great or small created a novel object there would be an infinite number of worlds, and selves.

On this view, change is adventitious, with properties by and large inessential to the object. For example, bi-pedalism is part of a common definition of human but not the use or disuse of the limbs, in walking, running, sleeping or paralysis. Certain properties are critical or paradigmatic to the object, as to its sameness, such as, in man, consciousness and personality. In substance ontology, change is not ingredient but something an object undergoes, an activity or inactivity that is extraneous. Substance ontology internalizes perceptual objects, carving up and populating the psyche with "logical solids" similar to those in the world. The ontology is reinforced by speculation on the timeless, eternal or changeless ideas of Platonic thought, or the cognitive stasis of a set of "bloodless categories", or a lifeless past forever fixed in time. Energy is the stuff of particles but Feeling is conceived as supplemental. Ultimately, substance is a composite of a host of elements all the way down to basic entities that are themselves compounds of external relations.

In contrast, the dynamic of process theory, or becoming, which is fleeting and unobservable, must explain why the world seems to contain innumerable substances, how they are stabilized and how they appear independent of the observer, since the cognitive process that underlies substance has no perceptible correlates. Becoming condenses into the affective content of the objects of thought and perception. Yet we feel and observe an arousal of ideas by emotion, or emotion as an incitement to action. This accentuates the energetics of emotion at the expense of conceptual form. All mental contents and objects are categorical frames of Feeling. From an internalist perspective, categories are consolidations of embedded phases. From a physiological standpoint, they are segregated nodes or vibratory levels that enfold primordial Feeling. The Absolute of process ontology is relational and dynamic, like the *pratītya-samutpāda* of Buddhist metaphysics (Brown, 1999) or contemporary string theory. Object-formation is a becoming into being; there are no solid things. A rock, Whitehead wrote, is a mass of raging particles. Process and substance ontology are complementary. The "rock bottom" foundation of the world is not at all rock-like; it is purely relational.

Objects *seem* to change before our eyes and effect change on other objects. In process theory, change is in the becoming of actualized categories. Every change is a novel recurrence, and every recurrence is a changed world. This resolves a long-standing problem in causal theory, how a cause is carried into an effect. The disappearance of the cause in the effect represents the perishing of the past as it is overlapped and replaced by another world. The impression that change occurs across states rather than within them is due to the "invisibility" of becoming at the interior of a forming object. The idea of change in external passage conforms to common sense but an account of change in process is deep and counter-intuitive. Thus, a major difference in the ontology of process and substance theory is the locus of change, either in the becoming of an object or its causal transmission to another object.

Feeling and Emotion

In a complex organism such as a person, energy is active at multiple levels, from body cells and intra-cellular elements, to anisotropic Feeling in brain, beginning with instinctual drives and satisfactions and leading to conceptual-feelings and intentional aims. Moods are manifestations of non-local Feeling.[18] Some theorists postulate unconscious emotions and/or conflicts between them, but how can such emotions be identified if they are unconscious other than as states of tension, friction, anxiety, uncertainty, blockage, hesitation or an obstructed inclination in some direction. Unconscious emotion may be like spontaneous or automatic action that must occur before its value can be identified. Courage might be an example of an unconscious impulse or emotion. Affect tends to be diminished in rational concepts. In external objects, it is often imperceptible or seems projected by the observer.

Feeling sequesters at the posterior limit of the mental state. Concentration at the onset enhances Feeling in drive and primal Will, which dissipates in affect-neutral objects. The more intense a desire, the closer it is to drive; the less intense, the closer it is to objects. The transition from a concentration at the unconscious core to affect-neutral diversity at the conscious surface reflects the allocation of Feeling from unity to multiplicity. Feeling is not diminished but is divided and muted by allocation as it is colored by diversity.

Desire is intermediate in the derivation, relinquished at the outer limit of the intentional aim. In the partition to desire, sexual and hunger drive moderate when satiety is achieved. Intensity in the pre-object is emptied in its progression from inception to termination, in the succession of phases, in the "derailments" and distillations. Like a torrent that loses force in tributaries, the mental landscape is enriched in rational thinking as affect is impoverished. Conceptual-feelings or affect-ideas nourished by a common source become the dry shoals of conscious particulars. A wave of Feeling shaped to a configuration by neuronal populations that correspond to instinctual drive is analogous to a mental category. Brain-activity is the dynamic of cell populations;

---

[18] The idea of emotion as a secondary interpretation of bodily changes or as originating in brain probably needs rethinking in light of this discussion. Energy and directional Feeling inhabit the body in liver cells as well as neurons, so that an energic theory of mind/brain is continuous with bodily feeling.

mental-activity is their conceptual frame. A category segments a traveling wave.

Let me close with a word on the deep relation of category and Feeling. In one sense, Feeling as energy or process is aligned with brain activity, while a category is a mental construct, so the relation of the two is equivalent to that of mind to brain. I have argued that process is a traveling wave of whole–part shifts constrained by unconscious patterns of the prior state, of beliefs, values and habits, with the resultant configuration sculpted by sensation to model the external world. If the process can be depicted as a series of nested whole–part shifts, what is the nature and origin of a category and how might it correspond to whole–part process in brain? If the whole–part transition is central, and if wholes can be said to correspond to categories and parts to members, the whole-to-part transition would map to the elicitation of acts, ideas, emotions and objects. That is brain process and the mental state can be described in terms of whole/part, context/item or fractal-like transformations.[19]

A related question concerns the relation of a category to its members, especially with regard to Feeling. A category is an abstract "structure" that encloses a set of shared, tacit or implicit possibilities. A category member that becomes conscious or explicit aborts the category for the particular, which then becomes a sub-category for a range of potential members. For example, once *dog* is elicited from the category *animals*, the prior category dissolves and the new one (dogs) appears. The antecedent category is replaced by the consequent one (dog), which includes canines. Process is forward-looking. However, any content can be assigned to some background category; one can withdraw to earlier, deeper phases, but in mind-active, items forecast sub-ordinate members, not super-ordinate categories.

Feeling is an ineffable pattern of vibratory activity in the brain that satisfies this description, for it remains indefinite until the category resolves. Once an idea, emotion or object clarifies, the antecedent category is left behind. The potential for unrealized particulars now belongs to the category of the elicited content. Desire has a multitude of possible objects that narrow down in love, then partition to affection, friendship, compassion and so on. When the reverse occurs, e.g. going from interest or friendship to love, the replacing state revives the ante-

---

[19] Other studies that refer to ground/figure, surround/center, frame/content and so on seem to be groping toward the same description, as well as more generalized accounts of individuation, specification or differentiation, though none of these accounts have mapped cognition to process in relation to a concept of the mind/brain state.

cedent category to which feeling (interest) is subordinate. All concepts and attendant affects follow this pattern. With the implementation of a drive, potential members of the category are eliminated.[20] Similarly, the category of desire is forfeit when its object resolves, i.e. the potential for an object is abandoned when one impulse clarifies. Conversely, desire that begins with interest transitions to what is prior in the mental state. Since drive and desire are aroused in every mental state, the passage from interest to love replaces the more superficial (object-close) category of interest with the deeper (self-close) category of love.

---

[20] Displacement of drive is well described in the ethological literature. With blockage in drive-expression, a return to core potential elicits substitute behaviors.

Chapter Six

Thinking

Bodily Decrepitude Is Wisdom.
—W.B. Yeats (1932)

Introduction

Typically in mind/brain investigations, cognitive systems and subcomponents are isolated for more precise study, and treated as interactive modules that can be severed without loss of explanatory power The thesis of this chapter, indeed of all my writings, is that such relations are *internal,* i.e. components do not come together by external contacts but appear as isolates by virtue of a uniform process ingredient in each of them. Diversity is elaborated, not assembled. The alternative to modularity is not holism; it is a diachronic process in which a cascade of fractal-like but qualitative whole–part shifts gives (the appearance of) separation and multiplicity. On these grounds, an understanding of one mental activity involves the whole of cognition. With regard to thinking, this includes, *inter alia,* perception, feeling, memory, agency and the self.

As to the relation of self to thought, one can ask, can a subject exist without some mode of thought going on? In human thought, is there a self, an "I", without thought, or thought without an "I"? Is the self, the "I", or the core, the "me", a *source* of thought, a part of thought or extrinsic to it? Is there a progression from self to thought and, if so, is it reciprocal? Do self and thought interact, i.e. does the self shape thought, and the reverse? Are they mediated by common or separate mechanisms? Is thought primarily a state, process, content or faculty? Does the outcome of thinking stand apart from the process through which it develops or does it include antecedent and unconscious phases? Can thinking occur without thought, as with a dormant or unresolved problem or an erratic stream of images? Is thinking going on all the time or is it intermittent? Is there unconscious thought? What is day-dreaming? Is the spectrum of thought a continuum, say from dream or imagination to logic, or are there different substrates for each form? What is the relevance of thought disorders such as psychosis, paranoia or hypnotic states? Other aspects could be emphasized, such as effort, volition, spontaneity, propositional content, intentionality, habit and innovation, pragmatics and creativity. Thinking embraces so many differing states or activities that one is hard-pressed to come up with a definition, even a description that does justice to the diversity.

Thought and Perception

Thinking subsumes many modes of thought: visual, verbal and gestural (in the deaf); imaginative, fantastic, symbolic, animistic, mystical, realistic and logical;[1] automatic and deliberative; incidental and obsessive, musical, mathematical and so on. With so many types of thought and an infinite number of contents, if we take only the content we miss the process of thinking. The same problem occurs in perception, though perception is not usually confused with thought since it seems to begin with a mind-independent object, while thought arises out of the shadows and remains in the mind. Objects seem real and persistent; thoughts are often vague and evanescent. Thought is something the subject creates; objects come from outside and, unlike thoughts, are part of an extended space. The distinction of thought and perception entails a different relation to the self. In perception, the self is a passive spectator, in thinking it is a more active participant. The *cogito* can be interpreted as "I think" — a self that thinks — but also as thought occurs, with the "I" thought up in the act of thinking. Most people believe they think up their own thoughts but not that they think up the objects they perceive — except for this writer, psychotics and an occasional philosopher.

Privacy in thought and publicity in perception begin at opposite poles, one in conscious or unconscious mind, the other in the world. An object is the external source or instigator of perception; thought is the outcome of thinking. Since content in perception appears extrinsic to perceiving, i.e. the object and the act of perceiving are conceived as distinct, interest is on features and mechanisms common to all objects, e.g. color, shape, movement. We can look at a field, a tree in the field, a branch, a leaf or an owl on the branch, so what counts is the process through which objects are perceived, not the objects themselves. Perception is adventitious to its object since the same object can, presumably, be perceived by others, while thinking is foundational to individual thought. This accounts for the tendency to replace the process of thought by the thought-content. Systems such as logic presume

---

[1] Those who hold logic as the paradigm of thought might recall Heidegger's (1959) comment that "logic relieves us of the need for any troublesome inquiry into the essence of thinking."

the autonomy of content without incorporating anterior process; those that do invoke antecedents, such as psychoanalysis, assume early experience exerts an effect across decades. This works well enough in ordinary dialogue, perhaps also for psychotherapy, but not for theory of mind.

The distinction of thought and perception is radically altered if object-perception incorporates the phases out of which it develops, such as (the substrates of) dream and imagination, i.e. if segments in the mental state that subtend dream and imagery correspond to intermediate phases in perception. This does not mean that a dream is lurking inside an object but that a failure to realize external perception, or an enhancement of preliminary cognition, allows penultimate phases to come to the fore, either as images in the context of object-perception or as final actualities. In imagination, these phases are active but embedded in object-perception. In dream, they are endpoints. In immediate perception, they transform to objects. The substrate of the image, not the image, passes on to perception. The consequence is that the object is the aim of process not the starting point,[2] with pre-terminal phases — dream, imagery — not ancillary to perception or memory but forms of thinking, i.e. thoughts (concepts) that become objects.

The conscious imagination is a mix of object and image. Imagery is usually not as vivid as dream, and objects do not attract focal attention. Dream is inner perception, i.e. perception that terminates prior to sculpting. In dream, endogenous imagery does not transition to an object. The outer world is not realized, the imagery propagates independent of adaptive constraints, and the self is passive, at times even a victim (the seed of paranoia) to its own imagery. Hallucination, dream, imagery are a melange of perception, thought and memory. Perception and memory also overlap in the concept of the trace. Memory is generally conceived as the residue of an object that is "stored" or reactivated, the two distinguished by the pastness of memory, the lack of externalization and the mixture with thought. Thought and memory are not confused with perception since the latter becomes extrinsic as it conforms to the environment.

Comment on Language

If thinking is defined as problem-solving, the genesis of human thought can be found in the most primitive organisms. However, there is little question but that human thought is closely bound to language, so much so that linguistic analysis often substitutes for an account of

---

[2] The idea that perception is active and productive goes back at least to Bergson (1896/1959).

thinking. But how is this relation to be understood? Does language constitute or implement thought? Does it generate visual as well as verbal images?[3] To what extent does intuition or spatial thinking depend on language? Is inner speech the equivalent of verbal thought and, if so, is a description of inner speech a description of thinking? How does inner speech differ from thinking aloud, argumentation or conversation? Clearly, language is responsible for human thought. In the first year of life, thought-development is roughly the same in human infant and chimpanzee, but with the advent of language, the human infant shows rapid growth in thought leaving the chimp far behind.

The relation of language to thought is of critical importance but, however dependent thought is on language, this chapter is concerned with a uniform *process of thinking* to which meaning, syntax, lexical relations are subordinate. The problem is not dissimilar from that of feeling, though language is central to thinking, while feeling is conceived as a separate activity, even in opposition to thought. The similarity lies in the fact that as thought is not bound to sub-functions of language, neither is it bound to specific emotions, which are segments in a derivation (Brown 2012a). The focus of this paper is the genesis and role of verbal (and other forms of) imagery, with the goal to understand the pattern and dynamic of the thought process and how diverse systems individuate.

Feeling

Since Plato, thought and feeling, reason and appetite, have been sharply distinguished, with thinking attributed to the "higher centers" of the brain and emotion relegated to the basement. However, a theory of thought, or any aspect of cognition, requires a concept of feeling and how feeling relates to concepts. In my view, thought and feeling are not dissociable. Feeling is the dynamic in concept-formation, and concepts embody an affective tone. Thought begins in instinctual drive, with categories of hunger and sexuality such as edible objects, predators and mate selection that, in humans, underlie the specification of object- and lexical-concepts (Fig. 6.1). The drive-category combines a feeling and a pre-object. This construct develops to a phase of conceptual-feeling, e.g. desire, at which point, as a unified configuration, feeling is enfolded by

---

[3] I recall a seminar with George Miller, who endured a discussion of language and the propositional basis of visual imagery until finally he asked in frustration, what is the propositional content of a phantom limb?

and innervates the concept. Conceptual-feeling empowered by drive-categories is foundational to thought and the conscious self.[4]

Mental states in which feelings like fear or anger seem independent of thought, or states in which reason and abstract thought seem independent of emotion, are differences of degree, not kind. The category enfolding an emotion establishes its object, direction and distinction from other emotions. In human mind, affect drained from ideas appears in preference, import or valuation. The feeling in abstract concepts or logical propositions obtains in their selection or valuation, in protest or affirmation, and in the often aggressive need to persuade. To take content (substance) in thought as independent of feeling, say a proposition, uncouples the content from its formative dynamic. The distinction owes to the fact that feeling is less pronounced in certain ideas than in their covert generation. Content that is relatively "affect-free" is extracted from affectively-charged antecedents. I have argued that the *being* of substance (concept, object) actualizes the *becoming* of process (feeling, action). Idea and feeling are unified at their onset, then diverge. Feeling is identified with process, content with substance, one obviating the other. The psychotherapist who listens to a statement and then asks how you feel about it is not, contrary to practice, inviting a reaction, but extracting feeling submerged in the process through which the statement actualizes.

Intentionality

Intentionality — the aboutness of consciousness — is synonymous with thought, including desire, which is necessarily about something, an experience, a problem, a feeling, an object or a memory. Judgment and valuation are usually construed as secondary to consciousness of intentional content. The sequence is presumed to go from being conscious of something, or thinking of something, to an appraisal of its value or truth, to action based on that judgment. Microgenetic theory reverses the sequence (Fig. 6.1). Accordingly, feeling begins with drive, the intensity of which is muted by unconscious belief, value, preference or disposition. Nascent feelings individuate to concepts which, as conceptual-feelings, pass to external objects. Value accompanies the object outward as worth. The actualization of content cuts off feeling in its formative arc.

---

[4] As is well known, Wittgenstein held that instinct was prior to reason.

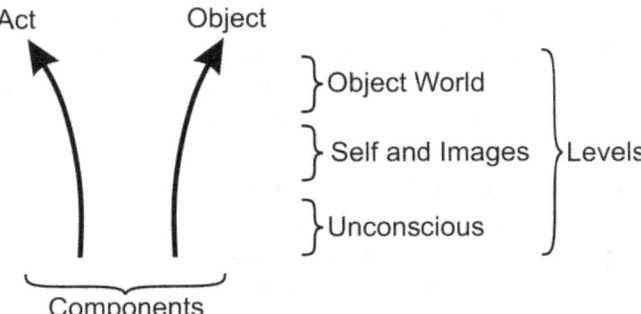

Fig. 6.1: Action and perception retrace patterns in forebrain evolution. A single traversal constitutes the micro-temporal structure of the mind/brain state. A parallel derivation of acts and objects recurs in a fraction of a second. The progression is unidirectional from past to present and from mind to world, leading from an unconscious core through intermediate phases to the external world. Sensation is applied primarily at the distal phase. In humans, the production and comprehension of language are grafted to the action and perception limbs of this system.

The aboutness of the intentional is one aspect of value, which applies to a positive or negative direction in thought or emotion. Is there a difference if consciousness is about something, values something or attends to something? What is the difference between aboutness and interest? The content to which thought or consciousness is directed is an outcome of interest, and interest, no matter how dry or rational, is still a manifestation of valuation. We do not think of, or look at, something and then decide if we are interested. Interest drives the looking beforehand. If we see or think about something, and subsequently develop interest in what we have seen or thought, the interest is not an addition to the prior state simply because it follows it. Rather, interest grows as an augmentation of bypassed segments in the original cognition that were gravid with value and undisclosed meaning. Aboutness is a trajectory from self to object or idea, not a particular content or unique relation to self.[5] Reason attains detachment in abandoning the trail of the intentional. A simple statement that seems affect-free such as "grass is green" is not independent of feeling, but incorporates implicit valuation, minimally in the demarcation of the content from other possible statements. Valuation is a mitigation of drive-energy by way of the derivation of unconscious need to conscious wish.

---

[5] In my view, the psychology of intentionality (Brentano, 1874/1973) has been hijacked by language philosophy, which takes the content of thought, such as a statement, as a mind-independent object to which a consensual appraisal or truth judgment can be applied (e.g. Davidson, 1980).

A description of intentionality in which aboutness is couched in terms of positive or negative, or approach and avoidance, i.e. an act that is toward or away from something, implies value in the choice and extends thought into like or dislike, i.e. desire and the host of related feelings it subsumes. The relative neutrality of an intentional content, say a proposition, can, if affect is subdued, shift to what is believed or not believed, what is feared or hoped for. There is a fluid transition from the unconscious root of the intentional to conscious intention, i.e. from need to want, from the "I must have (avoid, etc.)" to the "I have (know, etc.)". Willful intention becomes rational intentionality when impartiality replaces necessity. A change in direction or intensity gives different emotions, not more intense expressions of the same feeling.

Unconscious Thought[6]

How is the self (as well as thought, reason, feeling) related to action that is spontaneous or automatic or, for that matter, to action that is voluntary or deliberate? Is a thought that is suspended in the transition to automaticity re-engaged in volition? Actions such as walking and speaking can lose spontaneity. Does the return of deliberation for such acts require unconscious process to be filtered or molded into consciousness? What would be the nature of the unconscious phase? This is the thrust of Wittgenstein's claim that if a lion could speak, we would not understand it.

When deliberation becomes automatic, the unconscious choice is fixed in skill, technique or tacit knowledge. Several notions of the unconscious are implied: in one, a repository of beliefs, values, images and symbols, memories and other potential contents of thought, much of which is based on the experience and interpretation of dream; another related to processes that are part of the animal inheritance and never become conscious; another concerns events of the first few years of life that are irretrievable; yet another is related to skill and learned behavior, whether knitting, chess or grammar; and still another similar concept relates to procedural memory, which is like an unconscious skill such as riding a bicycle or learning a game of cards, usually contrasted with declarative memory, which is conscious and reflective.[7]

---

[6] Some philosophers, e.g. Searle, dismiss unconscious thought as physiology. But this is also true for conscious thought, except that subjective experience is added. If both conscious and unconscious thought are physiology, the difference is that unconscious thought is not conscious, which after all is the basis for designating it as unconscious in the first place.

[7] The distinction of procedural and declarative, or implicit and explicit, is relevant to the discussion of tacit knowing on the one hand and deliberation on the other.

We assume unconscious skills or memories are the foundation of conscious thought but, in microgenetic theory, unconscious traits combine with drive and early experience to form a core that undergoes further individuation.

We do not as often think of skill in relation to perception though such skills as mirror-reading, visual constancies, auditory categorical perception, eye-hand coordination and pattern detection pass from conscious learning to unconscious knowing. We speak of the "trained eye" of the scientist who knows what to look for or the detective who sees clues where others are baffled, but all of us navigate the environment each moment with little thought to much that is happening around us. Implicit perceptual judgments are continuously made with little if any conscious thought. Habits also fall into this category. The unconscious phase in thought can be embedded in consciousness or a terminus in dream. The transition to a further endpoint generates an image in the context of an object perception or as a final actuality in dream.

In sum, unconscious presupposition — a rudiment of drive muted in the specification to desire — passes through child-like or dream-like cognition (including imagery, primitive thought or animism) to actualize in objects, images or rational thinking. Phases submerged in the transit recur in a fluid exchange from depth to surface.[8] When extrinsic constraints are reduced from a failure to model the external world — dream, sensory deprivation — or overcome by intrinsic pressure, antecedents actualize as endpoints. With normal sensory adaptation a reversible withdrawal occurs to earlier phases of introspection, choice, imagery, fantasy, creative thought and mysticism. In pathological thinking, the preliminary is enhanced in *the context of a normal object-development.*

Self and Thought

Thinking is something the self does, and thought (desire, etc.) is something the self has. Thinking or desiring — the verb — is an activity of the self; thought or desire — the noun — is its outcome as an adjunct or possession. Thoughts can be discarded or forgotten, but the self is the person, not something the person produces; the self does not have a self. Different accounts of moral responsibility treat the self as an accumulation of selves over the lifespan, the sum or average of its acts or the self of the final conscious state. A self that is the outcome of prior states — primarily the immediately preceding one — is still an individu-

---

[8] The older literature beginning with the Würzburg school (see Humphrey, 1963) and Schilder (1935) took up the problem of images as outcomes of thought or aids to thinking.

ality, not a multiplicity. We recognize a mean or average self, i.e. the core that we identify with character, but the conscious or empirical self, unlike the core, adapts to exigencies of the moment.[9]

The conscious self is nearly-identical across contiguous states, in contrast to thought, which is constantly changing. While a change in the self leads to a change in thought, a change in thought is not a change in the self unless the self is overcome, in which case the self is transformed. If the change is temporary, the core recurs. Madness, passion, religious conversion, obsession, relinquishment, addiction or other states and pathologies can radically alter the self, such that we say "he is not himself" or "I no longer recognize her". The possibility of dramatic change is ever-present but for the most part we are "stuck" with the selves we are.[10]

The relation of self to desire (or thought) is less pronounced than that of agent to action, except that thought stays behind while acts externalize. Thought is expressed in motor or verbal acts, but it also externalizes in perception, as hallucination. In the former, the sense of volition is retained, even growing stronger. In the latter, thoughts are like objects to which the self is passive.[11] A powerful thought or desire may usurp agency to such an extent that other thoughts are subordinated or eliminated. The slave to desire or one in thrall to a given thought relinquishes choice, open-mindedness and restraint. At other times, thoughts, fancies, hopes range freely while the self feels trapped or condemned. Except for obsession, hypnotic states, abject submission or perseveration and inability to deviate from habitual lines of ideation, and excepting creative thought, which often comes unsolicited, thought is usually felt as active, at times as induced or guided by subliminal forces.

An intrusion or obsession can take on a thing-like quality as an isolate in the stream of thinking. In schizophrenia, thoughts objectify; a thought (voice) may even command the self to act in a certain way. The self is a victim to its own thoughts, which are attributed to god, the

---

[9] Those who think they have discovered a distinction between a core and conscious self should read Kant on the distinction of an empirical self and self *an sich*; also James (1890), and my work over 20 years ago (Brown, 1991).
[10] Identity of the self is a complex problem that Hume deferred to later philosophy. James (1890) offered some valuable ideas.
[11] Agency and activity on the motor side of inner speech (verbal imagery) complement passivity and receptiveness on the perceptual side. The receptiveness of late phases in object-formation accompanies the detachment of objects and hallucinations as mental contents seem to impinge from outside (cf. Brown, 2009, for how this may occur).

devil or another person. Since hallucination derives from the self[12] — where else could it arise? — it must represent some facet of the self's belief system. A schizophrenic person who hears an alien voice will attribute his thoughts to an imaginary other. The paranoid person hears a voice that reinforces or complements his thoughts. In rare cases, body parts detach and no longer belong to the self (anosagnosia), or a limb may interfere with the self's own actions (alien hand).

Stability and Identity

The self feels an instigator of thought and action, while thoughts and acts are felt as effects. The direction goes from self to thought with an outcome in action (Fig. 6.2). The feeling of direction is deeply ingrained in the psyche. However, one can ask if the self, as an early phase in object-formation, is deposited as a type of thought, first unconscious or subliminal, then supraliminal, to then give rise to conscious acts and objects. The identity or unity of the self, which is not felt for thought, is forged of implicit beliefs, values and experiential memory. Nor is agency felt for thought in the same way. Identity and agency are incompatible with the episodic nature of thought compared to the relative constancy of the core.

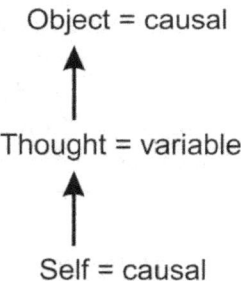

Fig. 6.2: Self and object recur more or less the same. The self recurs by way of internal constraints, the object by way of sensory constraints to achieve the stability, concreteness and self-similarity essential to causation. Thought is variable. Only in special instances does it attain a replication comparable to self and object, at which time it too can be felt as caused or causal.

The unconscious core and the self-in-consciousness recur largely unchanged, in contrast to the mutability of thought and desire. The

---

[12] A former student, Steve Levick (1986), was able to mitigate lateralized auditory hallucinations in schizophrenics with an ear plug on one side. He speculated on hemispheric imbalance; I wondered if the effect was achieved by participating in the patient's delusional system, i.e. that the voices were coming from outside.

feeling of agency owes to the stability of the self in comparison to the dynamic of thought. The earlier locus of the self and the near-identity of recurrence do not usually allow for an opposition of selves. While some may assume this is the basis of conflict, i.e. competing self-concepts, it points to conceptual feeling where choice, thus indecision, is possible. In everyday life we recognize a diversity among innumerable thoughts in the same self, most of which arise, fade and never recur. For an inference of causation, however, recurrent thoughts must have some modicum of stability. Thoughts that achieve stability, e.g. ruminations, obsessions, *idée fixe*, even propositions, become temporal objects with a thing-like quality.

Agent-causation is the causal effectuation of the self on mental contents, including processes that implement action. Psychic-causation is the causal role of mind in general (feeling, reason, thought). Thoughts, desires, feelings can grow stronger but, except for certain attributes, e.g. courage, a self does not intensify. A special vividness of an object would not necessarily provoke a change in the self, though it might arouse greater sensitivity, as in poetry, joy or melancholia. The intensity of thought over a brief duration, such as rage or intoxication, can replace a stability of mood achieved over many replications and lead to actions that would not otherwise occur. Thought is irrational when the self is irrational, but irrational thought occurs in rational minds. The self is not identical to its thoughts, but every thought is a glimpse into the self. It is natural to say a person (self) acts in a bizarre or irrational way, but not that an irrational thought causes the action.[13] Only when thoughts recur and gather intensity can the self be said to be one with its thoughts.

The stability of self and object is established in recurrence. Thoughts, ideas, propositions remain in the mind as logical solids or externalize as consensual facts. The persistence of causal objects, or a belief in their substantiality, depends on the similarity of recurrence across states. The precision of replication owes to the physiology of the replication and the entity to be replicated, which can change, e.g. ice appears substantial, water is less so, and vapor least of all. From a process standpoint, a tree is not a stable entity "out there" in the world but a model that recurs, actually a category of like exemplars. The tree motionless in the field, shaking in the wind, bent over with snow, leaves falling, is the same tree, or rather the *category* of the tree, the members of which are sampled in a given perception. The category of a

---

[13] Some take the person to be more than the self, even to include the body or social interaction, but since everything including others and the body is in the mind of the observer, there is no justification for this distinction.

self also remains much the same over different states and conditions. In all instances (e.g. self, thought, object), stability is ingredient in the intuition of causal power. A self that recurs is apprehended as relatively stable and capable of agency. An external object that recurs is apprehended as substantial and causal. A thought that recurs takes on motivating power. Inner and outer objects are presumed to be potential causes or effects in ordinary causation, or in *causal persistence*, which is of greater relevance to the mental state (Russell, 1948; Brown, 1996).

The stability of the self can survive an occasional rupture, for example an episode of weakness in a person of courage, or an act of foolishness in someone who is prudent. Unless such occurrences are marked and frequent, we do not interpret them as signs of instability but as instances of human fallibility. For the most part, constancy is assigned to character, while deviation is interpreted as anomaly of circumstance, stress, depression, fear, alcohol or some other influence. The core is felt as persistent though it can be shaken and distorted. Self-identity is the inevitable outcome of the overlap of states, but it is also a coping strategy for the individual as well as the inference of mind in others. Identity depends on substantiality, which in turn depends on several factors, including the invisibility of the process through which the content develops, the categorical nature of content, and the belief that stability is real and non-illusory.

The categorical nature of objects[14] is evident when their defining or paradigmatic features are retained on different occasions. A bird that is dying or in flight is more or less the same bird that a moment ago was perched on a branch, but a sparrow that transforms to a crow is perceived as hallucination, dream or delusion, in any event, something thought up by the self, not a substantial existent. Solids are like waves in the ocean that are carved out and frozen. The belief in logical solids reinforces the feeling of a substantial, non-momentary self (Fig. 6.3).

On this view, thought is an outcome of a mental state that provides reasons for actions but does not cause them, i.e. thinking is exculpatory, not causal. Deliberation gives the illusion of choice in the suspension of automatic action and thus a feeling of freedom. Paradoxically, the feeling of agency is accentuated by fostering indecision. Deliberation accentuates volition in an act that might otherwise be impulsive. The thought of an action, or the conscious desire to act, is not a necessary preliminary to bringing the act about. It is more natural to say, "I decided" or "I acted", implying that the self is an agent, than to say

---

[14] Buddhist philosophy holds that all objects and part-objects are categories, each feature being a category to subordinate items down to particles or vibrations.

"thought decided" or "reason caused an action". It is possible to say "thinking on the problem made me act in such a way", but the operative word is "me" or the core.

All action occurs in the body with a secondary effect on objects. Objects leave the mind to become independent, but actions remain in the body and, except for unusual dissociations, are felt to belong to the self. Instinct and drive are bound up with the "body image", "schema" or bodily awareness. The "me" of the core is embodied, the conscious "I" less so.

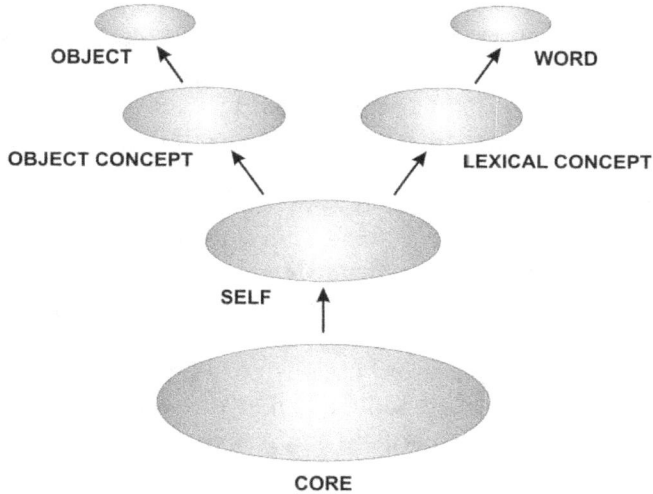

Fig. 6.3 is a sketch of the succession of categories from core to self to object and lexical concepts and finally to objects and words. The progression from unconscious to conscious goes from the general to the particular but all contents—an actual endpoint or embedded phase—enfold a multiplicity of virtual members in a partition to definiteness. The transition from core to self to thought is replaced by a transition from core to self to action. The activation of core, then conscious self, at the base of every state gives the feeling of an agentive self behind thought and action.

In sum, the stability of a thing or mental content is a necessary component of its inferred causal status. Thoughts, images, dreams lack causal power because they are fleeting. But causation entails more than stability. A conscious thought or feeling cannot cause an act because it perishes as an outcome; it is replaced "from below"; it is ingredient in the state as an indivisible epoch. Within the state, a thought could be an effect of an earlier cause (phase) but not a cause of a later effect, since the causal entity would be the entire state, no segments within the state, and not thoughts, which are final actualities. This is why, following Dewey, theory should center on the mental state as a transition from potential to actual, which is closer to psychic reality than one from cause to effect.

Agency and Causation

Is thought expressive or causal? In psychic- or agent-causation, a self (or thought) is assumed to cause behavior. We tend to accept the idea that thought (reason, desire) is a causal product of the self, and that the self is the cause of action, but we are less convinced that thought is itself a cause of other thoughts or actions. Does one thought call up or cause another in a stimulus–response chain? In agent-causation, the issue goes beyond causation to freedom of choice. The feeling of volition, the consciousness of options and the decision to act, heighten the sense of agency (Chapter 9). But a self that is driven or constrained to implement an action can still be causal even if the sense of agency or voluntary choice is wanting. With consciousness there is always a self, and with a conscious self there is always choice even if the choices are implicit or unpleasant. Agency inheres in consciousness though the self can feel helpless or passive to circumstance.

The conscious self is felt to be the cause of action though we sense, vaguely, that the self is susceptible to unconscious influence. We have no direct knowledge of an unconscious cause, yet we assume the self has sources of which we are indirectly if at all aware. We are prepared to believe this is so, and speculate on unconscious beliefs, motivations and conflicts.[15] If the self arises out of an unconscious core, the core, not the conscious self, would be its cause. If the self causes action but is not the *primary* cause, that is, the self is an effect of anterior process, what is its cause? Is thought also an effect of this cause? As the parent transmits the grandparent's genes to the child, the unconscious motive is conveyed into conscious action. Is the self a way station in the conveyance of memory into perception, or the past into the present?

Thoughts come and go and the thinker is often at a loss to explain why some thoughts are conscious and not others. If a thought that demands action precedes the act, is it necessarily causal? If I think (decide, plan) to go to the store, does the prior state of mind — a moment or a month ago — cause the going to the store, or does a subsequent mental state replace a prior one? The acts of a lunatic seem driven by irrational thoughts, but one can as well say that thoughts

---

[15] Collingwood (1940) argued that conscious reason arises from unconscious presuppositions. So too, presumably, does the self. The self that decides, chooses and is felt as the initiator of an action is an outcome of antecedents that include predispositions or proclivities guided by unconscious values and beliefs, attitudinal tendencies that are part of character.

rational and irrational are attributes of the self to which agency is assigned. Does thinking anticipate action as an excuse or justification for what is unconsciously "decided", i.e. does reasoning provide reasons, not causes? Does thought support, fill or rationalize unconscious bias?

We are guided by unconscious tendencies in the same sense that a dextral bias in the tonic neck reflex, the "fencing" position of the newborn, predicts later right-handedness. An orientation bias in the axial plane forecasts a bias in the distal innervation. We do not assume a gradual development from axial to asymmetric action in the distal musculature—though no doubt it occurs, in maturation and in the mental state—but we do speculate on the passage of unconscious to conscious thought.

A Causal Self?

The sequence from conscious self to thought is partly the basis for inferring a causal relation, but does not carry into the sequence from conscious thought to action. Action may be essential to the genesis of volition but not to the later feeling of agency. Indeed, the self that reflects on the possibility of agency exhibits, in the process of reflection, the agency it is deliberating.

As suggested by Freud and others, thought fills the delay before action, or is a surrogate for action or is trial action (Englefield, 1985; Brown, 1987). Memory can also be said to fill this delay. Both thinking and remembering are essential to agency. However, these activities go on in dream when the self is not an agent so what counts for agency is conscious thought or memory. If I think or remember to go to the store and then go, or decide not to go, the self is apprehended as an agent but thought and memory are not proximate causes, though they might be conduits. It is more accurate to say thought-imagery in a fully-unfolded state gives some feeling of volition.[16] When an explicit image in one state is revived *implicitly* in another one, i.e. embedded in the phase transition, it delimits and implements and to that extent determines the action. This implies that thought and action do not occur in the same state but are successive realizations, a state of thought giving way to a state of action (e.g. as in Fig.6.4, state B is replaced by state C).

Consciousness implies choice and decision. To think of going to the store is to entertain the possibility of not going, or going elsewhere, so the self, if causal, does not necessarily impel action but may inhibit

---

[16] The feeling of volition is distinct from agency. Volition is the feeling of choice and agency. Agency is the presumption that the self causes action. Free will is the volitional quality of choice.

action or lead to inaction or another action. Inaction is still action and a mode of causal persistence, like self-similar replication. A persistence that results from replacement can be construed as causal. The state is constrained to reproduce itself. Are constraints causes? Hume was uncertain. *Object-causation* – the impact of one object on another – is a theory of change, in mind and the world, though causation involves cognition if we ex-Hume arguments on the relation of psyche to necessity. Guyau argued (Michon *et al.*, 1988) that causation and the idea of the future develop in the child's reach for an object. Agent causation is the causal implementation by the self of bodily or vocal movement. A direct effect of the self on objects would be telekinesis. A succession of proximate events in mind is vital, though a causal effect of one mental state on its successor, or the successor on its predecessor, is uncertain.[17]

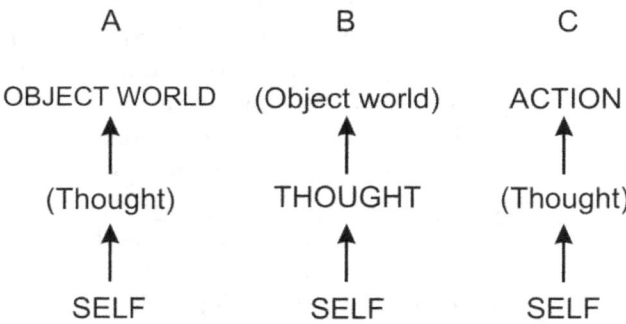

Fig. 6.4: Mental states A, B and C. At A, the self passes to the object world with thought *implicit* in the transition. This would be the case in spontaneous action or perception when the unconscious "choice" associated with thinking is buried in the act. At B, the self passes to *explicit* thought and the object world is implicit in the background. A thought actualizes as the dominant focus of the mental state, and adapts to the world by way of sensory constraints. These constraints are suspended in dream as thought (or perception) is liberated. At C, the self passes to action with thought implicit in the transition. If the explicit thought at B is implicit in the mental state at C, the self will feel that the thought empowers the act.

The question of whether the self causes thought or action depends on whether and how causal theory applies to a cascade of qualitative shifts within the mental state. Is there a causal progression from onset to termination in a transition that is timeless until completion? Does causation apply to a fractal-like succession of wholes to parts, or

---

[17] A careful reading of Brown (2010a) on serial order will show a legitimate possibility of backward causation.

categories to items, in which the parts are qualitatively unique and guided by inner and outer constraints? In the elicitation of part-categories from those of wider scope—think of the specification of animals to dogs, or dogs to terriers—the parts are both unique and general; they are themselves categories that do not sum to the antecedent whole, e.g. each partition encloses an infinite number and types of terriers, dogs and animals.

The *passage* from depth to surface in the mind/brain state is comparable to that in a fountain, the body of a river, even a ripple in a stream. It is not clear that one segment causes the next to occur. How does an earlier stretch of a river causes a later one, into which it will be transformed, i.e. the upstream becomes the downstream, it doesn't cause it? The slippage of before into after in a wave-like continuum is not causal *if the earlier becomes the later,* and the segments are not demarcated. A fundamental distinction of microgenesis with conventional theory is that, in the latter, a mechanism outputs to another mechanism while in microgenesis a wave-front is guided to actuality by constraints on developing form. If the traveling wave is carried through to the endpoint and undergoes transformation in passage by the influence of constraints, in what sense is this causal?

A related question is whether the shape of the river bank causes the shape of the river, or if the river carves out the shape of the bank. The impact of extrinsic stimuli on endogenous process, i.e. sensory constraints on the developing state, might be causal, but not the process-sequence on which it acts. There is also the matter of causal change in the replacement of epochal states. Since overlap is not temporal until a full cycle of existence is complete, any causal effect would be confronted with the non-temporality of replacement, i.e. the replaced segments are early in the state prior to actualization in time. Distal segments of the state are not overlapped, and achieve temporality when the epoch actualizes. The mind/brain state is a temporal epoch that perishes, so the genesis and completion of the state, viewed so to say from inside, do not occur in physical time as commonly understood. The epoch elaborates a bubble of the present about which past and future can be apprehended and, in this, creates an illusory subjective time in contrast to the physical passage of earlier and later.

To summarize the preceding deliberations, some problems for causation, taken up in the next chapter, are:

1. Before and after and the continuum of becoming.

2. The lack of demarcation of phase, and thus of cause and effect.

3. The becoming of cause into effect, i.e. the passage of earlier into later.

4. The epochal nature of the mind/brain state.

5. The concept of a causal sequence in a non-temporal context.

6. The extrinsic locus of constraints and the internal relations which they influence.

7. The perishing of actualities on realizing temporal existence, i.e. the object perishes prior to an effect on other objects.

8. Change in the generation of the state and the constraints on the derivation.

9. The replacement of proximal, not distal portions, of the state.

10. The dual time series: before/after within the state; past, present, future when the epoch is realized.

11. The non-existence of separate phases or segments in becoming.

The actualization of the mind/brain state and its replacement cannot be reduced to conventional accounts of cause and effect, for example to a sequence of causal pairs, though cognitive science tries mightily to do so. Process and becoming conform to a nested series of whole–part transitions of categories, for which causation is an inadequate theory. Further, it is the case that causation is beset by the very problems that are resolved in microgenesis, not the least of which is the importation of the past into the present, and the identity across boundaries. In my view, mereological theory is preferable to causal theory, but one in which wholes are potentials of becoming, not containers of parts, and parts are not constituents of wholes as compilations but possibilities of specification that do not exist *in statu nascendi*. Every part is a whole or category that either terminates as an actual object or leads to further partition. For these and other reasons, an explanation in terms of potential and actual is preferred to one in terms of cause and effect. The trick is to give scientific or mathematical validity to the shift from potential to actual without sacrificing quality for quantity.

Action, Perception and the Memorial Basis of Thought

Motility is often treated as the read-out of a keyboard or action modules in motor cortex activated by "central processors", with downstream discharge to brainstem and spinal cord. For microgenesis, action develops in parallel with perception "upward" over distributed systems in forebrain evolution. The system is hierarchic, with successive levels (e.g. Yakovlev, 1948) enfolding into axial and postural musculature, then distal asymmetric movement. Language production is grafted to this system. Like action, speech develops over a sequence of kinetic rhythms that begin with oscillators for the breath group and respiratory timing, then the speech melody and actualize in the fine temporal program of speech sounds. In line with the preceding discussion, the elicitation of a series of rhythmic levels in which harmonics develop out of more fundamental frequencies is comparable to the specification of categories to members, which then serve as categories for another round of specification.

The temporal lag in (visual) perception—minimally, the time for the transmission of a signal to retina, optic nerve, geniculate body, optic tract, visual cortex and, in the classical view, the assembly and recognition of the object—is a delay prior to actual perception that is necessarily off-line with unobservable nature.[18] This delay corresponds to one in voluntary action, e.g. pre-activation of up to 0.8 seconds prior to a conscious decision to act (Libet, 1999). In other evidence for pre-activation, the onset of a voluntary movement of a finger coincides with the peak of normal resting tremor, demonstrating that an oscillator underlies the conscious decision to act. Action and perception develop in parallel to conscious acts and objects that adapt to, or mirror, entities in the world. The findings support work long ago by Bernstein (1967) on motility, and Martin (1972) on speech, on the entrainment of oscillators or kinetic rhythms prior to the conscious perception of an object and prior to the conscious decision to act.

In the perceptual withdrawal of schizophrenia, thought mixed with recollection signals the onset of delusion which, like dream, is a mix of the two. The more image-like the content, the more thought-like the experience. The pre-activation of conscious voluntary action must

---

[18] The temporal lag alone shows the memorial basis of perception. Merleau-Ponty (1992) wrote, we remember objects into consciousness, and Whitehead (1921), "there is no essential reason why memory should not be raised to the vividness of the present fact." Recent studies by Eagleman (2011) on this topic, including those on the fluid duration of the now and virtual relation of self to motion and voluntary feeling, support microgenetic theory.

correspond with the perceptual lag, for otherwise action and perception would dissociate. This also shows the onset of the mind/brain state is no doubt tightly bound to the passage of physical nature but in consciousness acts and objects are virtual or phenomenal. Such observations support the conclusion that thought, memory, emotion and other "components" of cognition, though routinely demarcated, are in fact differing manifestations of a uniform process in which one mode is momentarily dominant, the varied implementations being just so many tributaries of the same stream.

The Relation of Memory to Thought[19]

The description, duration and theory of replacement of the mental state, the transmission over evolutionary growth planes, the passage from past to present—from the memorial to the perceptual—the sculpting and externalization of image to object and the relation to subjective time, briefly touched on in this chapter but extensively discussed in prior works, are interpreted from a single coherent theoretical standpoint. The theory applies within and across different modes of cognition and provides an account of everyday experience as well as aberrations in pathology.

There are many works on thinking, older standards such as Bartlett (1958) and Bruner *et al.* (1956) and recent ones such as Kahneman (2011), but none to my knowledge that deal with the continuum from one content or mode of thought to another, the relation of thought to image and object or its relation to the self, agency and causation. There are also few works—Heidegger is an example—in which thinking is related to theory of mind. Writings tend to be devoted to skill, logic, modularity, artificial intelligence, cognitivist and linguistic models,

---

[19] The biological correlates of memory have been well studied, though the resultant division of memory into components vitiates the import of research findings. In contrast, the correlates of thinking are obscure, as are those of most other mental events. Presumably, these correlates are neuronal populations that, by way of recruitment and synaptic strength, configure a population-dynamic over successive phases. The configurations probably develop as a field effect (Eccles, 1972) or traveling wave (e.g. Tucker, 2008) incorporating lines of development leading to memory, thought, imagery and emotion. The wave-like individuation is unidirectional from an archaic core to a conscious object, from unity and simplicity to complexity and diversity. Spatial fields associated with specific capacities, e.g. imagery and thought in egocentric space, lead to actions and objects in Euclidean space. A critical feature is the sustained specification (analysis) of wholes into parts, guided initially by the internal constraints of habit, value, belief and the immediately prior state, then by the external constraints of sensation (sculpting) on distal segments.

problem-solving and decision theory. Accounts of thinking tend to focus on the products of thought, in which the process of thinking is equivalent to its exemplifications.

The micro-temporal development of the mental state is uniform with a bias toward one outcome or the other. Thinking is just a way of describing mental process when the outcome is thought. Remembering, imaging and emoting are ways of describing the process with a different outcome. The relation of thought to memory is evident when memory is productive or when thought is reproductive. The relation of thought to visual and auditory imagery and inner speech (verbal imagery) is evident in introspection. The relation to emotion appears in the partition of conceptual-feeling to affect-ideas (Chapter 5). Thought can be infused with emotion, emotion may call up thought. Thinking is the pattern of the process with a bias to imagery when memory and feeling are in abeyance.

Obviously, thought develops on a bedrock of experience and learning, whether mathematical, scientific or creative, most of which is unconscious in the form of skill, habit and the like. Conscious thought and memory are a mix of the familiar and the novel. Memory is imported to the present as thought, with an emphasis on the past. An intentional memory is transitional to a productive thought. Thought is for the possible, memory for the irrevocable. Memory in the service of thought is "working memory", the technical term for holding a memory in the mind to aid in problem-solving. This view is an artifact of research in which memory is distinct from thought, in spite of the obvious fact that experience is a mix. This mix of thought and memory is illustrated in dream when the content is not fully individuated (Fig. 6.5).

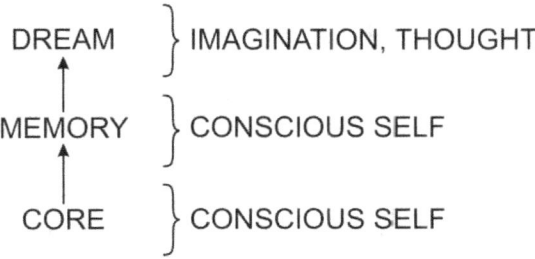

Fig. 6.5: Without the sensory constraints that would transform the phase to objects, the microgeny actualizes in dream at a phase of imagination and thought.

Is dream a memory transformed to a thought, or a thought confounded with a memory? Often it is not possible, on waking from a dream, to say which part was thought and which part memory. One can ask, is a memory that is condensed, fused or revived in symbolic form still a memory? Does the "dream-work" signal a transition to thinking? One speaks of a memory as incomplete, distorted or derailed, but the deviation can be a sign of thinking. The inability to distinguish memory and thought is not evident during the dream but on waking.

In dream, the past is occurring *now* without comparison to prior experience. The lack of familiarity for dream content raises the possibility that *the conscious sense of familiarity is critical in the distinction of memory and thought. A memory not apprehended as familiar will be taken for a thought, e.g. repeating one's self or appropriating the work of others without being aware of it. A novel thought or experience that feels familiar, as in déjà vu, will be taken for a memory.*

If it is the case that the sense of familiarity determines whether a content is a thought or a memory, what accounts for familiarity? The arousal of prior experience entails an *implicit* comparison that should occur in every mental state; otherwise, as in the film *Groundhog Day*, each day — each moment — would be a replica of the preceding one. However, the sense of familiarity is unusual for everyday events. Familiarity informs the subject the content is memory, not thought. When I drink my coffee in the morning, why do I rarely think of the act as repetitive? It is unlikely that familiarity arises from the memorial nature of the content; in any event, the explanation is circular.

Is familiarity a product of memory or an aid to its identification? The prominence of familiarity in *deja vu* where thought is confused with memory, and the absence of *deja vu* for most memories, implies that more than familiarity is needed.[20] This additional something, I would hold, is the sense of *pastness* that inheres in memory but is foreign to thought. A feeling of pastness in thought would make a thought seem as if it was thought up before, in other words, that it is a memory. The feeling of pastness owes to an earlier locus in the developing mental state, which specifies the past of earlier experience, even childhood, in relation to the immediacy of perception. We know that older brain systems, such as limbic formation, associated with memory and emotion, are enjoined early in the state. Later phases extend memorial content into thought, then into object-perception. The past is alive when the pre-terminal recurs or undergoes transformation to the verbal or visual imagination (Fig. 6.6).

---

[20] In *déjà vu*, there is also a feeling of the uncanny, which may result from the clash of past and present for the same content, briefly exposing the fact that we remember objects into perception.

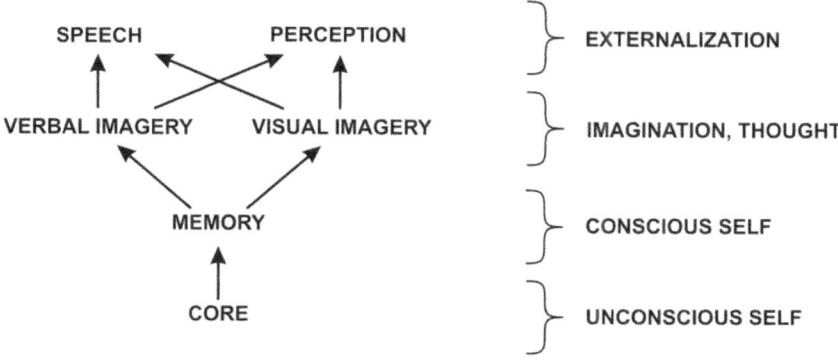

Fig. 6.6: Sketch of the development of personal and experiential memory into two main paths: (1) verbal imagery and speech; (2) visual imagination and (visual) perception. Visual images can be described or implemented in speech, and verbal imagery is the ground of speech perception. Emotion undergoes a similar partition. Action (limb, vocal) and perception externalize in an adaptation to the world, in contrast to dream which, without sensory modulation, terminates prematurely as "inner perception".

We understand the common-sense distinction of thought and memory, in daily life with forgetfulness, and in diseases that affect memory. We know people with unexceptional or poor memory who excel in thinking, and the reverse.[21] Conscious memory can erode while skill remains intact. I have seen many brain-injured or demented patients with severe amnesia and preserved skills, for example, adroit surgeons with a profound memory loss who can perform complex operations but forget them a moment after they are completed. Thinking may be relatively spared until old learning is affected. However, in everyday life, thought and memory are intertwined. One says, I was thinking about a walk yesterday or a vacation last year. Thinking about the past is reflecting on memory, or memory mixed with thought. To think is to entertain novel ideas, not to resurrect old knowledge, though this can pass for thinking.

---

[21] The ability to retain long texts or new languages on brief exposure, such as learning the bible or teachings in Sanskrit, are examples of prodigious memory, but in memory prodigies there is often a weakness—perhaps a reciprocal deficiency—of conceptual thought (Luria, 1968). The story by Borges, *Funes the Memorius*, makes this point with considerable insight. Unselective recall and poverty of concepts are limiting factors in thinking. Another is forgetting what is irrelevant—or not reviving it, for this is necessary to recall the essential.

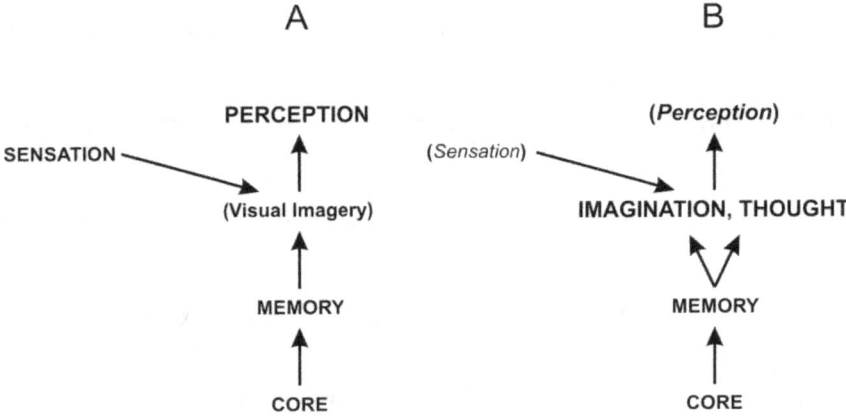

Fig. 6.7: (A) shows development through the substrates of visual imagery to perception. An internal image (brackets) does not develop. Instead, the substrate that mediates imagery adapts to the world by sensory constraints. This is ordinary perception without introspection. (B) shows the heightening of conscious thought and memory with a reduction of sensory modulation or when the pressure of antecedent phases drives mental content to verbal or visual thought (introspection). Objects appear through distant sculpting but the focus is on memory or thought, and the outer world (brackets) is in the background.

To say I remember a walk or vacation seems to limit the recall to an isolated event in a relatively brief duration, since a memory that constantly recurs is a sort of rumination, if not a pathology, that in its digression is equivalent to thought. A distinction could be made between the spontaneous recurrence of a memory, such as that evoked by a taste or encounter, and the voluntary reminiscence of past experience, which is like thinking in which memory substitutes for thought (Fig.6.7).

The conclusion is that *if the content of thought is felt as past and familiar, the state is recollection; if the content is felt as novel and goal-directed, it is thought; and if the content is felt as spontaneous and/or random, it is imagination. A conscious memory is thought (imagery) guided by intrinsic constraints. A conscious object is thought (imagery) guided by extrinsic constraints. A relaxation of inner or outer constraints allows different forms of imagination to surface.*

Memory is the starting point of thought, but thinking, because it develops out of memory, can evoke past experience. A grandmaster who plays multiple games of blindfold chess evokes the strategy of each game or the pattern of thought in order to reconstruct the location of the pieces. Here, memory is activated by thought, while the chess player draws on the memory of combinations to plan a possible move. A dissociation of thought and memory points to the foundational status of memory in relation to thought, which depends on learned skills and tacit knowledge (Polanyi, 1973). Thought deviates from memory in the exploration of its infrastructure when constraints on accurate recall are lifted to uncover what is concealed or suppressed (Fig. 6.7). For example, in writing his great poem, Wordsworth recalled a voyage through the Alps, including "associations" latent in memory, e.g. category members, the propagation of object-property relations, as well as the signification, emotion, excitement and fatigue, the sounds and sights of nature, in a word the full richness buried in the experience and decanted in composition.

Conscious memory that is not entirely skill-based is episodic,[22] i.e. a particular event or event-series at a specific time, even if it is not dated, and even when the memory is replayed as a topic for thought. The dynamic of thought entails a more persistent engagement. Thought can lead anywhere; memory follows a "script". A thought that recurs over time with little change is a memory. In rote learning, a lengthy series of events, music, poetry, whole books may be recalled, but ordinary memory tends to be restricted to a given category to which specific events are related. In contrast, thought by its very nature is divergent and productive. An alteration of memory is often a substitution or elaboration, not just an omission and any deviation in recall is a kind of thought. Conversely, weakness in conceptualization returns the individual to a familiar outcome that, if repetitious or habitual, is closer to memory than thought.

The ability to activate old knowledge and overly-learned, largely unconscious skills mediated early in the state is the axis on which thought develops. An inability to bring a past experience into consciousness is a failure of thought as well as memory. The inability to consciously recall (think of) a specific event is an inability to achieve definiteness. The more precise the to-be-remembered item, the more vulnerable to forgetting, e.g. names, words. Thinking, however, circumvents specificity and makes use of roundabout strategies. *What is lost in a disorder of memory, at least in the early stages, is the ability to think of (reflect on) a specific memory, while unconscious skills that are antecedent to conscious thought, or on which thought depends, may be relatively unaffected.*

The physical past of memory is theoretically fixed but the psychic past is unstable. In pathology, false memory or confabulation may accompany amnesia. Few can sustain a remembrance in consciousness

---

[22] This does not refer to the technical distinction of episodic and semantic memory. In my view, semantic memory refers to categories of like-episodes the particulars of which have mostly been forgotten.

without transformation. Memory is not a copy but a re-creation. To maintain a memory in consciousness is to revive it recurrently through a dynamic self and the novelty of thought it distributes. But it is difficult to hold in consciousness *any content;* memory, thought or emotion. In some ways, an unremitting focus on a specific content is comparable to the attention to a perceptual feature, in which subtle eye movements (micro-saccades) are needed to avoid the refractoriness that occurs when the gaze is fixed. When eye movements are paralyzed, objects disappear. It is equally difficult to concentrate or meditate on a static content, a memory, a thought, a mantra or scene from the past, since any deviation of the "inner gaze" entails an emendation of content. There are, of course, those who live in the past, or those for whom open-mindedness in thought is subsidiary to fixed opinion. Yet, conscious memory is not consciousness of an unexpurgated trace. It is a reflection on the past, on what, why and how things were or might have been. This is memory as a goal, not memory implicit in thought.

The distinction of memory and thought is emphatic when rote learning is contrasted with problem-solving[23] in the artificial setting of an experimental study, but the recall of facts is rarely the machine-like retrieval from a hypothetical store that is the dogma of research studies. Yet, on the basis of such studies, memory is isolated as a special faculty and divided into a variety of forms, including stages in retrieval, e.g. short- and long-term, working, iconic, modal-specific and so on. To take these divisions as separate operations or mechanisms without clear lines of transition among them, or without a relation to other capacities or modes of cognition, obscures the dynamic and creative quality of recall—not to mention its relation to thought, emotion, etc. Synchronic division is not complemented by diachronic relatedness even if flux is closer than interaction to what is psychologically real.

In the reproduction of a past event, inexactness is thought's entry point. False memories are thoughts in disguise where the individual may not even be aware of the inaccuracy. For a person to know a memory is inexact implies a standard or template to which it can appeal. This standard is the unconscious construct, or potential, configured by the experience out of which the conscious event is specified. An implicit comparison also occurs in thinking. How otherwise would we know that an idea is original, or that ideas and memories are our own?

---

[23] The problem with "problem-solving" is that it combines human thought with adaptive behavior in spiders, birds, perhaps paramecia, a catch-all concept that incorporates the entire evolutionary scale with little sensitivity to the uniqueness of human thought.

William James wondered if he was getting mixed up in other people's dreams. How do we know we are the proprietors of our thoughts and memories and not just parroting other people? Well, often we do, but less often do we admit it. Knowing is deeper than thinking, in thought and in memory, and in both the parts individuate the whole behind them. The potential realized in thought or recollection provides an intuition of ownership, originality and recurrence, as well as certainty and uncertainty.

Chapter Seven

Novelty and Causation

From an internal perspective,
causation is successive replacement.
The cup and the grasping hand
are replaced by the cup-grasping hand.

Introduction

The contemporary literature on mental causation attempts to tease apart psychic and physical entities — substances, states, properties, mechanisms, processes — as to causal efficacy in the relation of mind to brain, and the reverse, i.e. one- or two-way interaction, as well as within and across brain and mental states (Kim, 1993). The causal role of physical properties on or within mind, the reducibility or identity of mind with brain, its elimination or dependence or causal effectuation, reflect — directly or in response to — a metaphysical agenda in which causation is prior to explanation. The guiding assumption is that physical (physiological) causation in brain is an instance of universal causation in nature. A fundamental question is whether brain function is, in fact, causal and, if so, is mind explicable in terms of causal process in the mental state or in relation to the brain state. Does causation apply to mental properties and, if not, how else can the relation of mind to brain be explained?

The idea that brain events cause mental events involves a temporal step from a past brain state to a present mental state, or from a present brain state to a future mental state.[1] If mental properties coincide with brain properties, i.e. are extensible, causation would be possible across segments within mental states. Such properties would allow two-way causation as well as causal effectuation from one mental state or activity to another. Some problems for any theory of mind/brain, but especially one that arises in a substance metaphysics, are the contextuality and co-dependence of mind and brain, the interaction of logical

---

[1] A posterior cause or future effect that cannot exist leads to problematic claims for simultaneity in causation. Microgenetic theory attempts to resolve this problem.

solids within and across mind and brain, and the lack not only of a concept of mind/brain but a theory of consciousness, self, feeling, thought, memory and the relation of conscious to unconscious process.

So far, these are difficulties that no amount of philosophical speculation has been able to overcome. What is needed is a theory of the mind/brain state, of passage from onset to termination within the state, of succession across states, and without a precondition of causality, which only puts blinders on imaginative thought. Absent such theory, reduction, elimination, identity and so on are back-door solutions that abolish the problem instead of solving it. Given the facility with which mind is abolished for the sake of a causal metaphysics, one might expect a greater effort to explain the powerful feeling of self and agency. These may be illusions, but without an explanation of how they come about, no amount of argumentation will displace them. A causal metaphysics is so ingrained that an escape from determinism, particularly as to free will, hangs on speculations on emergence (below), microtubules, quantum indeterminacy and the like, which may liberate mind from causation but do not explain agency.

While the claims of this narrative reflect a failure to ground speculation in theory of brain and mental state, substance metaphysics and systems theory (Bertalanffy, 1968) entail multiple layers from physics to consciousness—emergent or causal—that develop out of, or are explicable in terms of, scientific concepts, such as whole–part relations. However, before addressing the problem of inter-level transition, process ontology is focused on the transition *within* levels, the pattern of which, I would argue, is the specification of parts out of wholes, not the reverse, with wholes as potentials, not containers, and parts as potentials for further partitions. The possibility of emergence in the transition from lower to higher begins as a within-level individuation; qualitative "fractals" are novel occasions not constituents of prior wholes. The occurrence of common patterns in mind/brain process could facilitate cross-level mapping.

Microgenesis and Causal Theory

Microgenesis takes a different perspective on these issues, having developed as a theory to account for language and cognitive errors with focal brain lesion. The theory represents a radical departure from accounts of brain and/or mental activity as causal, or of mental activity as causal by way of reduction or correspondence to brain function. The widespread belief that brain activity is governed by causal laws, to which mind is reducible or identical, is so pervasive that the theory cannot be ignored, even by arguments that offer viable alternatives (e.g. Filk and von Muller, 2009; Atmanspacher and Filk, 2011). A central difference with microgenesis is that the objects of philosophical speculation are isolated from their antecedents, which excludes from analysis the formative history of mental content.[2] The outcome of an act of cognition—an image, an action, a proposition—is effectively split off from its momentary ancestry and treated as an instigator, a concomitant or a resultant of causal process. In contrast, process (microgenetic) theory entails, in my view, that the conscious contents (events, properties) extracted from a micro-temporal sequence perish without effecting causal (or any) change. The construal of contents or events as causal, directly or through brain, conflicts with the idea that the *transition* to an act, object or utterance is the primary vehicle of change, not conscious implementation.

Another distinction concerns the tendency in philosophy to simplify the input and output limbs of a mental state, probably a legacy of computational models that postulated "central processors" surrounded by input/output connections. The result is a fictitious concept of perception and action as an incoming and outgoing extrinsic to philosophical interest, not as hierarchic systems through which mental contents or events are realized. The sources and goals of sensorimotor systems cannot be dismissed as problems for neurology or matters of hardware. Preparatory phases are ingredient in the processual structure that lays down introspective content. A perception is a series of formative phases, not an assembly of inputs. A thought is a preterminal object; an object is a specification of antecedents—thoughts, concepts. Actions unfold to cortical endpoints in parallel with object-formation. *In sum, acts and objects are not assembled by external contacts, but are hierarchical systems instantiated in mind/brain as outgrowths of a momentary ancestry. The state is epochal, implementing tacit knowledge and the experiential history of the individual in the traversal of virtual or actualized phases.*

In microgenesis, the process of momentary act and object development is a *becoming* over successive categories. Whole–part or category–member transition is a continuous *becoming-into-being*. The iterated partition of wholes (categories) generates the *being* that the becoming becomes. An epoch of becoming incorporates a category of phases; an actuality enfolds the partitions of all phases in becoming. This difference from philosophical accounts is apparent in phenomena such as the shift from memory image to object or the reverse, or problems such as

---

[2] For example, Mele (1987) notes that irrationality poses a difficult problem for philosophers who base arguments on logical or rational thought. In microgenetic theory, a layer of pre- or para-logical mentality is traversed in every occasion of thought or perception.

volitional feeling for imagery or agency for non-existent phenomena, such as phantom limbs. These are not aberrations. They uncover micro-temporal process within consciousness and provide essential data for mind/brain theory. An endogenous series of psychic phases leads from self to act, image or object.[3] The formation of the world, and action in the world, represent the furthermost limit in a continuous sheet of mentality from the onset of the mind/brain state to its terminus as the external world.

On this way of thinking, mental contents — not just thoughts or feelings but limb movements and external objects — are instantiations of mind/brain states. For an event-ontology, objects, contents, properties, any isolates in thought, are terminal or pre-terminal endpoints that perish and are replaced; in a word, they are artificial slices without causal power. Contents are not independent of the epochs in which they are embedded. Ideas, images, emotions, propositions actualize and vanish. A thought, a judgment, doesn't do anything; change occurs in its replacement. The becoming of the mind/brain state is a before/after series until it actualizes, at which point a present (being, substance or the category of categories) is created.[4] The process of elicitation of parts from categories is the becoming of the state, substance is the category through which the object (state) becomes what it is. The initiation and termination of the mind/brain state, from inner to outer, from subjective to objective, constitute an actualization that is a diachronic, modular and overlapping whole. Contents are *continua* with antecedents and consequents. This chapter argues that transition in the brain state maps to transition in the mental state, and that the alignment in mind and brain of qualitative change and novelty in whole–part shifts is incompatible with causal process (see related arguments in Atmanspacher and Filk, 2011).

If whole–part transit describes process in the mind/brain state, microgenesis is compatible with dual-aspect theory. Mental and brain states are hierarchically-ordered phase-transitions that individuate virtual or actual contents in a partition of qualitative fractals out of

---

[3] An action is a perceptual image generated by recurrent collaterals in the action discharge. The contribution to mind of action as motility was the topic of an exchange between James and Wundt. My position is that action contributes only the feeling of activity (the *Innervationsgefühle* that was the topic of debate), while other aspects of the experience of an action are perceptual.

[4] The notion of becoming-into-being, the simultaneity of the series, the before/after relation (becoming) and the elaboration on completion of a present (being), correspond with McTaggart's (1901) two series in time (Brown, 2010a).

neural configurations or mental categories.[5] There is much evidence for this in clinical data but the documentation of a similar process in brain requires a neuroscience not based on the synaptic model, e.g. flow diagrams, but on fields or wave-fronts (e.g. Eccles, 1970). Instead of the assembly of modular units in mind and/or brain, there is an iterated specification of potential to actual. The approach provides insights for subjective time-awareness that can resolve some difficulties in causal theory going back to the Cartesian concept of a material (causal) brain and an immaterial (non-causal) mind.

Whole-part fractionation has other features worth noting. To mention just one: if the core of a mind/brain state resolves out of a deeper whole on the order of a Jungian archetype or still more fundamental Absolute, this potential could give rise to a multiplicity of states. An intuition of this possibility leads to mystical ruminations on a deeper substrate out of which all minds devolve. Another feature concerns the possibility of emergence of consciousness through whole-part transition in the relation of early potential to distal actuality. If parts are *qualitative individuations*, each whole-part shift gives novel and unpredictable partitions that imply emergence for "higher" as well as "lower" levels. In brain, this goes from archaic to recent in evolution (sculpting is the paradigm); in mind, it goes from proto-mind to consciousness over phases. The concept of partitions as qualitative fractals presents difficulties for mereology, for causal theory and for the idea of consciousness as a result of complexity.

A final and critical consideration is that the continuity of the mind/brain state from onset to termination, and the common process of whole-part specification at successive levels, entails that rather than multiple processes at various stages in cognition, a single process, iterated at multiple levels, is applied to qualitatively novel partitions. This implies that the cognitive process involves a single configuration shaped by differing constraints at successive segments, much like a river, or a wave rolling to the shore, or a fountain, each subject to internal and external constraints; for example in the case of a river, the bank and bed. One would not want to say the upstream portion of a river causes the downstream portion, for it becomes the downstream portion; similarly, the lower portion of a fountain does not cause the upper portion which it becomes. Nor does the tidal ebb and flow of a wave readily admit to causal explanation. This way of thinking about the transmission of a configuration over levels shifts the debate on causality to whole-part transitions and the individuation of actualities out of potentials.

---

5   See MacLean (1991); Vandervert (1990).

Context of the Problem

A distinction can be made between *knowing what or why* and *knowing how*. The former—the capacity to know the *what or why* of a thought—is conscious knowing. The latter is the more or less automatic use of knowledge in producing behavior. The carpenter has a plan that may be conscious or unconscious as to *how* the work will be done, but he is conscious of *what* he is doing and why certain tools and procedures are selected even if unconscious skill enables a more or less direct translation of knowledge into action. The tacit becomes the explicit in conscious thought; thus, the distinction of reflection and habit, creativity and routine, choice and spontaneity.

The shift to and from the automatic and volitional is relevant in that actions unaccompanied by thought, though not freely chosen in the usual sense, appear to be causal outcomes, much as behavior in animals is purposeful but not volitional. The concern here is with the immediate ancestry of automatic action and conscious decision, which amounts to a description of the sequence that actualizes in behavior, whether automatic or deliberate. While the sequence from onset to termination, e.g. from unconscious to conscious, and from self to act or object, is a staged process, to maintain that the self or thought is the cause of an action is insufficient without filling-in the *how* of automatic process and the *what* of conscious choice. The mental state is an epoch that enfolds successive segments in passage. Since phases do not exist in time until the epoch is complete,[6] and since causation is a relation of past cause to present effect, or present cause to future effect, it is unclear how a succession of phases that is non-temporal or simultaneous can be causal.

The distinction of conscious choice and unconscious knowledge or skill, or dissociations of thought and action, is an old one, widely studied in pathological cases, particularly frontal lobe injury where the individual knows the difference between right and wrong, describes the right thing to do, exhibits conscious choice and rational decision, but still goes ahead with inappropriate acts, whether of violence (infrequent) or social impropriety (common), such as fabrication, public urination or tactless joking. The person will give a full description of a rational course of action, yet engages with awareness, even self-criticism, in behavior acknowledged as inappropriate (Pribram and Luria, 1973; Isaacson and Spear, 1982), admitting after the act that it was wrong, at times with regret. In memory, this relates to implicit and

---

[6] For Whitehead, one complete cycle of becoming (concrescence) is necessary for a thing to exist in time. See Wallack (1980) on epochal theory in process metaphysics.

explicit; in thinking, to automatic and volitional, or spontaneous and deliberate. One clinical approach treats the aberrant as a disruption or loss of verbal regulation by inner speech (Luria, 1962), but this hypothesis ignores a literature on defects of inner speech *without* prominent effects on behavior. Verbal regulation takes the description of behavior, or the conjunction of two behaviors — altered inner speech and lack of self-control — and assumes a causal relation.

Knowing the difference between right and wrong is critical in law and the assignment of blame and responsibility but all of us, from time to time, act in a way that is foolish, improper, even illegal, where impropriety, risk, satisfaction and judgment of success figure differently in the decision. Excuses, justifications and self-recriminations after the act imply conscious judgment, but what is most striking is the "disconnect" of reason and action, or more precisely, the fact that the impersonality of rational thought is trumped by need, whim or impulse. We say a person whose actions are directed by reason would not act in such a way, but we know that consciousness of what is right is not always a brake on conduct. Such individuals, if asked, would say their behavior is irresponsible. Indeed, much of what passes between people is not motivated by reason but by emotion and egoism with opinions formed before facts are digested. Reason is an exception, even if people tend to (habitually) behave in a way that is reasonable and in conformance with societal norms.

As to the distinction of knowing what or why and knowing how, if someone lifts their hand and is asked *what* he is doing, or *why* he is doing it, he could say, "I wanted to be noticed, salute, ask a question, reach for something, wave goodbye", in other words, reasons are offered for the action, which are assumed to be its cause. If asked *how* he lifted his hand, he would probably have no clue other than to describe the action-sequence or repeat why the movement occurred, such that the *how* of lifting is a result of the *why*, i.e. the self's decision to move, or he would follow some philosophers and defer the response to future neuroscience.

Observations on the difference of why and how concern the role of reason in the mental life and its relation to the implementation or inhibition of desire. The ability to distinguish right and wrong, or a failure to consider the consequences of an act, or indifference to its outcome, do not necessarily lead to good or bad actions. A suspension of conscious judgment would not necessarily result in an even chance an act will be good or bad or, given the primacy of the self-preservative instincts and their derivations, a greater likelihood of malicious action, since the recurrence of character, need and habit in relation to changing circumstance are the primary determinants of action and thought.

Character may not have good reasons for bad acts, which require good explanations, while good acts do not. The distinction of right and wrong differs from the nature of reason and its effect on action, since reason can justify almost any act. The moral sense relates to values instilled early in life and reinforced by experience, punishment and rules of conduct. Rational decision includes the role of judgment, *inter alia*, as an impersonal act.

In sum, the common belief is that rational acts are caused by reasoned judgments, but actions that reflect reason in decision-making more often are driven (constrained) by implicit beliefs, values and inclinations, not rational choice. When reason is a predecessor, inner speech (verbal thought) forestalls action and often depletes it of vigor and immediacy.

Rational Thought

If we set aside the specific content of thought, the first question is what is reason or rational thought? Is reason a particular use of inner speech and, if so, how is inner speech or verbal thought modified for rational purposes? After all, inner speech can be imaginative, poetic, meandering, decisive, inductive or deductive, synthetic or analytic. It can follow one line of deliberation or consider different points of view. It can pursue one option or several with or without a conclusion. It can be productive or reflective, engaging novel thought or reminiscence. Reason would seem to search for truth or reconcile options but it can also fabricate and dissemble. Reason engages inner speech to *adapt* to conditions — internal or external — close to real-world events. One can say that *practical reason is the adaptation in thought of need to circumstance.* The adaptive nature of reason is closer to speech and action, and differs from creative thought, which is less adaptive, or adapts to (creates) a reality closer to the inner life. The mode of reasoning differs according to the aims of the thinker and the methods employed to achieve them. A fundamental pattern in all forms of reason concerns a part–whole or one–many relation, in subject–predicate or topic–action relations, in syllogistic reasoning, in metaphor, synecdoche, going from instance to generalization, or eliciting a conclusion from a convergence of data.

Is irrational thought the absence of reason or does it follow certain laws or patterns that complement and forecast rationality? The question, what is thinking, is prior to an account of specific modes or modalities of thought — metaphoric, imaginative, syncretic, etc.[7] Rational thinking includes remembrance, dream and creativity at one

---

7   The importance of anterior process is seen in developmental studies, such as the demonstration by Vygotsky (1962) that verbal thought in maturity is an outcome of the internalization of egocentric speech in children.

extreme, formal logic at the other, and habit and reasonableness at the mean. One is governed by the laws of imagery, symbolism and a relative disregard of convention; the other by routine, conformance, opportunism, expediency, social expectations and norms, while logic involves specific rules that aim for mathematical precision in which adaptation to a particular realm of thought is shared by a group of like-minded others.

Creativity impels early thought to the fore, such as metaphor. There is greater subjectivity and, in art, a greater deference to the sensibilities of others than to objective fact. Ordinary reason conforms to social convention and common sense, while the conventions of logic, in language and mathematics, obey a set of formulaic rules. Probably, with mathematical logic in mind, Whitehead argued that the harmony of logic lays upon the universe as an iron necessity. Yet he also believed that a natural principle of Creativity is impervious to logic. Subjective antecedents — creativity, imagination — have a degree of freedom from time; ordinary reasoning is time-bound to concrete practicalities, while logic is impersonal and "timeless".

Is Reason a Cause of Action?

The notion that thought causes or controls behavior is, from a psychological standpoint, equivalent to saying that inner speech or verbal imagery causes action. What is true for rational thought is common to all forms of thinking. In fact, actions that are impetuous or ill-conceived seem a more likely example of direct causation than those that follow rational deliberation. Irrational thought more often leads to impulse, while rational acts tend to be planned and carried out with care, at times with a considerable delay between thought and implementation.

If thought and action arise from a common infrastructure, an influence by way of connectivity across transitional phases could occur at any segment prior to and including consciousness. Among the possibilities are: a direct effect of thought on action, an indirect effect by way of the self, thought as a processing template for action and mental trials or conscious veto. Intricate thinking with a delay before action, such as planning a voyage or rehearsing a lecture, presume that thought itself is a kind of action — trial action — especially in the absence of overt movement. Most occasions of thought involve reflection on events in the past or what will happen in the future, "things to do", bodily concerns, none of which translate unambiguously into action. The problem decants to the possibility, refuted above, that an image — momentary or prolonged (recurrent) — causes a movement.[8]

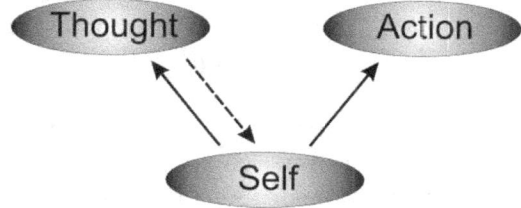

Fig. 7.1: If the self generates thought and action, for thought to cause action there must be a reciprocal effect on the self. Thought and action develop over a microtemporal infrastructure. Thought is in relation to perception and in parallel with the action-development. If a thought is an outcome that perishes, it cannot cause an action, which also has a hierarchic development. As shown in the figure, there would have to be "top-down", then "bottom-up" causation for this to work (see text).

If verbal thought stimulates, activates or channels behavior, reason could have causal efficacy by inhibition or excitation. Whether thought is rational, insane, imaginative or logical, it could influence behavior by the *elimination* of alternate routes.[9] Naturally, one would like grounds for asserting that reason has causal efficacy, not only because the possibility that reason or judgment is non-causal is a bitter pill to swallow, but because it raises equally complex issues as to the role of reason in the mental life, e.g. why reason evolved to formulate a choice, make a judgment, reach a decision or inform an agent, but not implement behavior. And if thought does not implement action, what is its source and what is the relation of action to thought?

A more felicitous explanation is an adaptation of the forming action to the guiding thought, initiated prior to consciousness and "fine-tuned" by constraints. More precisely, the actualization and replacement of thought *accompanies* the realization of action. When thinking is prominent, the action-development is muted. When action is prominent, thought is in the background.[10] This entails a shaping effect on

---

[8] We know that the thought of an action precedes conscious initiation, in some estimates (Libet, 1999) by up to 8/10ths of a second. A series of mental states, for the image or thought, for consciousness of initiation, for action, is more consistent with replacement (microgenetic) than direct causation.

[9] Hume was uncertain if constraints were causes. Of relevance to this discussion, Hume also wrote that reason can tell us what is true and what is false, but not what to do.

[10] The inhibition, substitution or restraint of action by thought, and the reverse, suggest common or aligned processes, but there are instances where action, usually axial, stimulates thinking. Children have difficulty

action at successive phases in thought. The transition from self to thought (inner speech) to vocalization would correspond to, thus entrain, the development of action, since even a part-act such as lifting the finger or saying a word rests on a micro-temporal process that lays down posture, gesture and orientation, not to mention concepts, feelings and the belief in a real object world.

This would be consistent with the claim that an action implemented by the self also generates inner speech as a way station in the derivation, or to delay behavior and sort out the most appropriate path to follow. On this interpretation, reason is a product of the self as trial action, much as inner speech and vocalization are presumed to be products of the self in agent-causation (Fig. 7.2). If the self is the proximate source of action, thought could influence action by constraints on correlated segments in the mind/brain state or penultimate segments in consciousness (Libet's veto), or by a reflexive effect on the self, which then induces the action. The latter could give the self a role in thought, presumably in action as well, but as a "top-down" effect it would entail the reversal of a presumed trajectory. In this instance, the effect of thought is indirect, for conscious judgment would "feed back" to—or be replaced by—a self that initiates the action, a scenario that is inconsistent with the direction of process in microgenesis.

In sum, a critique of causation and an account of process in the mind/brain state still require a depiction of the nature of thought and its relation to action. Microgenetic theory holds that states of thought do not induce action, but are replaced by ensuing states of action that are guided (constrained) by the overlap with prior states of thought, and the correspondence of phases in act- and object-development, which revive those pre-perceptual configurations that seem to cause behavior. The next section takes up the role of inner speech in thought, in speech and speech perception, and the critical fact that perception terminates in mind-independent objects, while action, except in pathological cases, does not fully detach from the mind but deposits in a bodily space anterior to that of objects. This allows the inner phases of thought to coincide with the terminal phase of action, reinforcing the illusion that thought causes action, and is not replaced by it.

---

holding still while they are talking. Adults often report—and this is my experience as well—that thinking is stimulated by walking, e.g. Brahms' description of composition aroused during walks in the countryside.

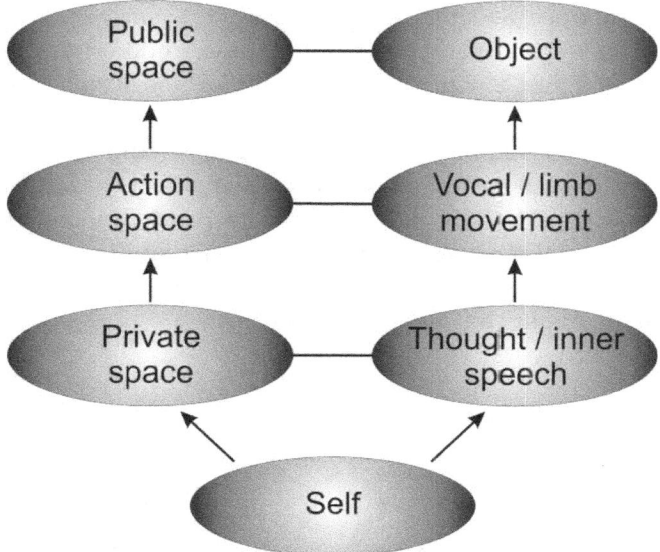

Fig. 7.2: Phases in the mind/brain state from self to act and object. The self is derived to thought (inner speech), which anticipates movement. Action-space is for discharge of vocal or bodily movement, and is preliminary to object-space. Phases in space-formation are aligned with phases in action-formation. Action deposits anterior to objects (Fig. 7.4). Movements in bodily space are felt as belonging to the agent, while objects detach in public space and are felt as independent.

Inner Speech and Verbal Thought

The inner speech that prefigures speech, writing or action can be thought-close or word-close; it can occur at the threshold of consciousness, or occupy consciousness as clear, precise and pre-vocal (Fig. 7.3). Inner speech has been referred to as the *preverbitum* when it spills into vocalization.[11] Subtle articulations have been demonstrated during inner speech with electrophysiological methods. But thought includes visual and verbal imagery; it may specify either type of image, propagate, recur and/or sink back into the shadows. The process recurs with words added, deleted or rearranged to express as closely as possible, i.e. adapt the thought behind them, to the external world.

In Fig. 7.3, verbal thought develops into action (speech) when the perceptual component is attenuated, i.e. incompletely revived (1). The active feeling of inner speech identifies the thought as voluntary, not hallucination or the voices of others. In creative thinking this active

---

[11] This idea is generally attributed to the Behaviorists, but it has a long history. For example, Bain (1868ed) wrote that "thinking is restrained speaking or acting."

quality is subdued and the person is passive to imagery. An attenuation of action (speech) and perception (vision, audition) is essential to conscious thought (2). Further attenuation leads to sleep. A relaxation of the impulse to act (3) accompanied by visual and verbal imagery, absent sensory constraints, gives dream or hallucination. The difference between dream and creative thought is the ability of the latter to move from inner to outer as the image adapts to reality. Inner speech and visual imagery pass to perception with external sculpting (4). If the subject is awake, verbal hallucination occurs when auditory sensation is insufficient to modify an endogenous percept.

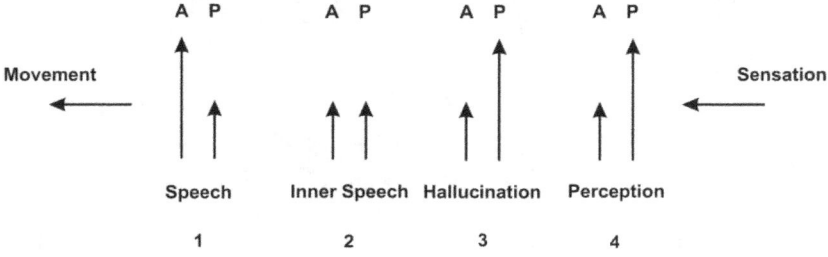

Fig. 7.3: A = action, P = perception. Inner speech is conceived as motor, especially when close to articulation but it is largely pre-perceptual (see text).

The same sequence no doubt occurs in other modes of thought, e.g. visualization in painting or modeling, where visual imagery accompanies or elicits movement. The motor quality of verbal thought—from vague presentiment to mental sentence—gives forward momentum to the verbal image. The effect is less prominent in creative thought, where the individual is receptive to his own imagery. Ideas and judgments not replaced by action remain intra-psychic, like phantom movements in the sphere of motility. These observations imply that a loss of immediacy in the delay of action is not due to a suspension of motility but a predominance of pre-perceptual ideation as a foretaste of actuality.

Further Comments on Thought and Action

In thought, pre-perceptions are accentuated but the outer world is retained. The thinker who is "lost in thought" is still aware of the world, even as objects recede into the background and images come to the fore. The pre-perceptual quality of thought coincides with phases in action, e.g. posture, sub-vocal and eye movements, gesture, automatic movements. A tacit "body language" accompanies each wave of thought, for example, a thinker who gazes outward, closes his eyes, scratches his head or shifts his posture. The action discharge exposes the infrastructure that anticipates the final phase of speech or writing. When verbal thought spills into speech, writing or other behaviors, actions materialize with the thoughts to which they are anchored.

Take the example of writing these words. Presently, I am relatively motionless on a bench in my garden. In writing, verbal thought develops as a gradual formulation that is transposed on paper. When an extended thought recurs, the content is written down to keep apace, or the thought is later revived. At times, I mouth inaudibly the words I am reading or writing.[12] At such times, verbal imagery seems to pass into speech and writing together, as thoughts flow to the vocal and limb musculature. Often, writing arouses ideas, the pen doing the work allotted to thought. The rare person conceives an entire work before composition; in others, composition facilitates thought. This reciprocity goes to the closeness of pre-perceptual imagery and incipient action.

In most instances, thought does not lead to action nor is action required, but rather, it propagates in the mind as diversion, self-exploration or play. In the absence of overt action, the potential for action, which is configured by thought, lacks the momentum to completion in relation to the thought it accompanies. A thought revived over many occasions may eventually lead to action. If implicit, the action feels spontaneous and the relation to thought is opaque. In other instances, the relation is clear, for example, an action that becomes automatic after a period of conscious thought and learning.

More generally, this line of inquiry impacts a theory of agent causation in the following ways:

1. If action does not originate in consciousness, conscious intent and content cannot cause action.

2. If the self instigates action that is guided by conscious or unconscious thought, how does thought, which is derived from the self, influence the self to act?

3. If an action is configured by memory, feeling and thought at multiple segments in its development, what is the locus, if there is one, of the causal relation?

4. If an action that accompanies a thought is replaced by a thought that accompanies an action, what is the causal status of replacement?

---

[12] The same mouthing occurs, sometimes audibly, in many concert pianists.

5. Some thoughts lead to behavior, most do not, or the thought *is* the behavior. How does thought, as a kind of act, relate to acts that express thought?

6. Implicit or explicit decision may guide or select a thought and an action. Is the selection of an action an act of thought?

7. Agency involves selection and implementation. If we understood the role of thought in action, we would still have to explain the effect of memory, feeling and expectation.

A plausible sequence is the following: in forestalling immediate (automatic) action, deliberation creates and fills a delay. Speech, writing, drawing and other motions punctuate and serially implement thought at penultimate phases in percept-formation, accompanied by a parallel development of incipient action. Action deposits in a space midway between mind and world, coincident with imagery but antecedent to the perceptual space of objects. Action belongs to the actor, like thought, while objects, though products of mind, detach from the observer and belong to the world. Inner speech or verbal thought, when biased by the action-development, leads to speech, writing or movement. It develops to hallucination or dream with a suspension of sensation, or to audition when the image serves as a template for perception.

Verbal thought (inner speech) and the action it accompanies, including speech, deposit at the same phase. Thought is felt as psychic; vocal or bodily action is felt as partly internal (belonging to the agent) and partly external (located in the world). Action contributes a feeling of activity but not *content*, except for the conceptual precursors that accompany pre-perceptual images. Constructs in thought, entirely perceptual, coincide with parallel phases in action as thought guides the emerging act. The infrastructure of action elaborates conceptual analogues of ideation generated in pre-perceptual imagery.

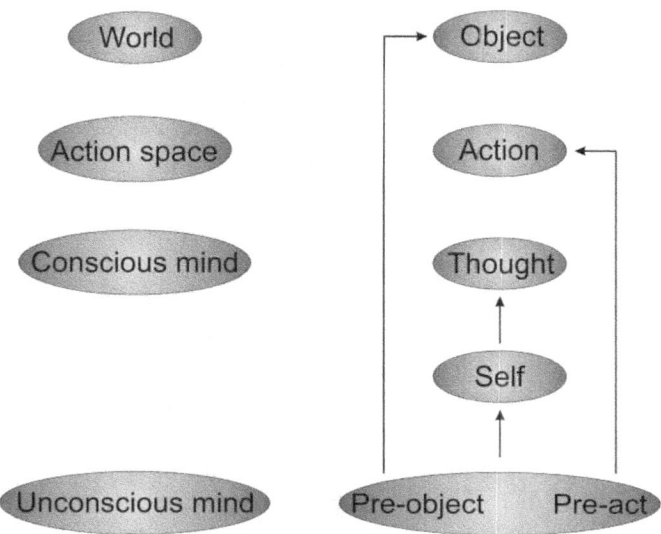

Fig. 7.4: The core gives rise to a combined pre-act/pre-object, then to the self at the floor of consciousness, then to ensuing phases. Action and perception diverge but undergo a parallel development. Action space is preliminary to the external space of perception, comparable to the limb-perimeter in infants and the space of the congenitally blind (see text).

Thought, Process and Agency

Reason can scrutinize a topic in thought when the goal is to understand, not act, though judgment and consciousness of choice develop together as adaptive solutions in non-habitual circumstances. A series of acts — verbal or motor — plays out in the mind instead of the world to reinforce commitment, search for outcomes or replace too-hasty decision, as well as reduce the impulse to act by incomplete analysis of antecedent potential. Specificity in action requires specificity in thought. The entire thought cannot be discharged at once but must be serially revived. Each possibility over a sequence of states resolves conscious choice with experience or tacit knowledge in a progressive adaptation to the physical or social surround. Thought is hindered only by the limits of the imagination, but actions have consequences. The commitment to action entails an adaptation to the world as well as to the whole of thought of which the action is a partial discharge. In this process, *the transition from wholes to parts, or from contexts to contents, is continuous from one phase to the next, incorporating the entire mental state, with world and mind as context and whole–part transitions the means of adaptation, e.g. sculpting, parcellation, evolutionary pruning.*

Every act involves a choice, mostly implicit. Minimally, a choice is a contrast with its negation. Habits and preferences are unconscious choices that become automatic by repetition. They recur as residues of conscious options long since discarded. Skill and technique are indelible patterns that begin with conscious effort. Options, decisions, the feeling of free will are only problematic when spontaneity becomes volitional and choice becomes explicit. Agency is empowered by ground-to-figure specification. In the passage to acts and objects, retardation of processing (neoteny) uncovers options in an inner space of introspection. Thought fills the delay brought on by prolongation (and elaboration) of partition. Postponement of decision allows options to surface. At times, a multiplicity of possibilities dilutes the affective tonality allotted to one line of thought, giving an ambiguity that can be corrosive to agency. Options are essential to agency, but too many options fosters uncertainty (Williams, 1981).

Affective conflict and/or an internal dialectic are foundational to agency. Conflict in the world resolves to predispositions in the mind that guide and constrain choices. Conscious options are derived from unconscious values and beliefs. Choice is not freedom; unlike action, it is relatively independent of objectivity. A self free to choose may not implement decision. A prisoner with diminished freedom can still think, feel, desire, decide, stand up, sit down, read a book, write a letter and, within these acts, choose the manner, mode and rate of activity, e.g. what book to read and how long, what to write and so on. Free will —consciousness of choice and the capacity to decide—should not be confused with freedom; e.g. to travel, engage in public discourse, political debate.

The inner constraints on thought, not the least of which are the skill, knowledge, character and needs of the individual, are no less decisive than circumstance. Conscious thought appears when routine is insufficient, or when action is blocked, or when novel solutions are required. Action has the same constraints, but must conform to the world. Free will and freedom are defined by the relative absence of limitation. On the psychic side, free will is limited by habit, obsession, perseveration, automatism, brainwashing, impoverished imagination, hypnosis, thought-control and, in animals and young children, spontaneity and drive. For freedom, the social and political environment is critical. The focal point of agency is the mediation of will and constraint. The focal point of freedom is the conformance of action and purpose. Free will and freedom—deliberation/decision, decision/action—are successive segments in a progressive objectification.

In sum, consciousness does not add choice to automaticity. Choice surfaces in the mining of the unconscious, not conscious manipulation. Delay in process (neoteny) exposes implicit or unconscious choice and the potential for other possibilities. An exposed potential reveals thought that is normally sub-

merged: metaphor, experiential memory, "lateral thinking", "dream-work". An alternate path in the whole–part transition penetrates consciousness as novelty. Agency is the instigation of choice in thought and its implementation in action, while the ability to speak or act freely is freedom. Thinking is a pre-object phase in perception that accompanies and constrains the action-development. When action is delayed – blocked, hesitant – thoughts arise as surrogates that guide the forming act to a conscious endpoint. The question is not why we have choice, or if choices are determined, but that consciousness, choice and thought are inevitable with a delay of action, since delay makes choice explicit, and choice and delay are inseparable.

Thought and Action: Summary and Conclusions

I have attempted to resolve the question, is thought a guide or determinant, and are reasons justifications, outcomes or agencies? Though reason and judgment are the primary topics of concern, from a psychological standpoint rationality is not especially relevant since the more general problem is the influence of thought on action, rational or otherwise. To say "reason (judgment) made me...", or "the voices made me..." refers to the self (the "me") using reasons as explications. The observation that thought gives rise to action, to writing, speech, painting or many other behaviors, is equally true for irrational thought, even if the content is inappropriate to the situation, as in psychosis, eccentricity, paralogic, ritual, ceremony or magical belief.

In rational thought, options adapt to the environment: physical, social, linguistic. Reason is close to the external. Distal segments of the mental state are further from need and closer to fact. The primary attribute of reason is the employment of words as verbal objects delimited in context by real-world constraints. Rational thoughts are verbal images close to objectification. Logical relations are mental correlates of object relations; grammatical relations in language are the equivalent of spatiotemporal relations in perception.[13] A logical statement is a surface derivation in which all contexts but the final words are excluded. A true or rational proposition is an internal representation that conforms to external conditions, in contrast to irrational thought in which the external is ignored or misinterpreted. The irrational precedes the rational in the evolution of mind; it is earlier in thought-production, closer to memory, to dream and fantasy, and fails to achieve a veridical endpoint.

---

13   Possibly, recursion in syntax evolves in relation to substrates of depth perception. Similarly, the virtual spatial image that resolves binocular disparity is analogous to the temporal image, the Now, that resolves the disparity of memory and actuality.

What follows are some variations on thought, action and causation in relation to microgenesis and state-replacement. According to this theory, *an act of thought is an act of perception with pre-perceptual phases arising as thought and nurtured in private space as precursors or consequences of action. Objects are externalized thoughts, thoughts are internalized percepts. A thought or action replaces itself in overlapping waves from unconscious inception to perceptual surface.*

1. In sum, the self individuates action in a traversal of thought, which guides or constrains the act to conform to rational or irrational ideas. Arguments for causation must be reconciled with the whole/part nature of the sequence.

2. Action has a hierarchic structure, passing to neocortex from a sub-cortical onset. Thought is a phase in the hierarchic structure of object-formation. Action and perception develop in parallel.

3. Every act of thought has a micro-temporal history that is part of a perceptual and action structure.

4. Thought arises in a partition to novelty subsequent to onset and prior to termination.

5. The essential steps are the transition of choice to decision and decision to action.

6. The relation of thought to action owes to shared properties in their actualization. Every act of thought evokes an implicit correlate in action.

7. If actions arise from the self, and the action-pattern is guided by thought, it must conform to processing in the mind/brain state, since conscious thinking cannot revert to the unconscious origins of the self and, like Lazarus of Bethany, regenerate a novel self infused by consequence phases.

8. Thought (perception) and action recur as parallel derivations of the self, with each component assimilating the other over the length of its derivation.

The priority of the self in action and thought and the attribution of agency to the self are less problematic than how the self knows what it is thinking if the thought appears at a subsequent phase. Conceivably, this can be explained by the epochal nature of the mind/brain state, in which all phases are simultaneous until the state actualizes, at which point the self is felt as spectator or agent to its derivations, i.e. passive to perception, active to action. If the basic pattern in mind/brain generates parts that are qualitatively unique, novelty would occur over all phases in the mind/brain state. A qualitative shift from potential to actual is not a causal sequence.

The relation of the self to pre-perception is comparable to its relation to object-perception. Consciousness of pre-perception is embedded in consciousness of objects (Fig. 6.4). The passive relation to objects accompanies their detachment. Agency depends on the intra-psychic locus of thought and the proximal space of action.[14] The relation of self to thought is less volitional than self to action and its processual nature. We feel our thoughts are our own; we think them up, even when, in creativity, dream and psychosis, they come to us as recipients. This is true for much of visual imagery, especially creative thought, but less so for inner speech, the motor side of which contributes to agency; otherwise, we would feel the insertion of ideas from others. This occurs in dream and psychosis. William James had the feeling in dream that he was getting mixed up in other people's thoughts.

The overlap for early phases in the mind/brain state embeds the prior state in the present one. The immediately preceding state is a template for the ensuing one, with each state an amalgam of the present and immediate past. The recurrence of proximal phases of experiential memory and tacit knowledge, and the perishing of the distal phase of objects, preserves memory and clears the way for novel percepts. The succession has the quality of a wave front, not a series of discrete events, less a chain-like sequence than a wave that rolls on the shore.

---

[14] The residue of agency for imagery is the idea of telekinesis for objects.

Chapter Eight

Certainty and Conviction

I act with complete certainty. But this certainty is my own.
—Wittgenstein

Introduction

This chapter explores the psychology of belief and the relation to truth and conviction from the perspective of microgenetic theory. The account of this topic in the philosophical literature is varied and extensive but for the most part belief is associated with truth, with intentionality and propositional attitudes, in which belief and proposition are identical or belief is directly accessed in a proposition. This tacit identification or direct access, or the imputation or a direct causal role of belief in behavior, effectively sidelines the micro-temporal transition — becoming — that leads from unconscious knowledge and experience, e.g. the memory "store", to conscious belief or statement.[1] A principle difficulty with this account is that truth is largely consensual and independent of belief, except for what the observer believes is true. The concept of knowledge as justified true belief would delimit knowing to rational judgment and carve out symbolism, value and creative thought from theory of knowledge. The notion that certainty and conviction in true belief represent degrees of confidence is surely misguided, since observation shows that strength of conviction is unrelated to truth or reason, and certainty is often provisional. Conviction is the emotional signature of the feeling of truth irrespective of whether or not the belief is true. The philosophical assumptions and the artifice built about them, in exaggerating the centrality of truth over adaptation, lead from the inner workings of psyche to an externalism in which belief and

---

[1] The contention of Bradley (1893, p. 224) that, "metaphysics has no direct interest in the origin of ideas, and its business is solely to examine their claim to be true", while perhaps applicable to the history of ideas, does not go to the momentary realization of statements. Yet the attitude is widespread in most philosophical disciplines at the expense of context, temporal relations, value and intent.

intentionality are inferred from behavior independent of the mental state and its diachronic context at successive segments in realization.

The condensation of belief to truth and propositional attitude owes to a copy theory similar to that for memory, storage and retrieval, which presumes that experience is revived unchanged into consciousness, a view that conforms to the computational theory of mind. The approach has little support in clinical study for it treats varieties of irrationality or false belief as ignorance or aberration, avoiding the process through which implicit belief becomes conscious and propositional contents are instantiated. The goal is to isolate philosophy as a formal system such as logic or mathematics, independent of the vagaries of cognition and individual consciousness, insensitive to context and oblivious to the psychological validity of its claims.[2] The norms of philosophical argument insulate the topic from external critique and reinforce the notion that psychological data are subordinate to issues of philosophical concern.

I will argue that belief is informed and reinforced by knowledge, with truth largely the reconciliation of need and adaptation. Conviction is a measure of certainty in belief independent of truth. A belief is not equivalent to a true statement for belief can be open, detached, dogmatic and fixed, rational, whether right or wrong, delusional, often to the point of intolerance, where protestation leads nowhere and refutation is useless. Belief as true knowledge owns facts that, for the believer, warrant conviction, while a psychology of belief engages patterns of thought—subject/predicate relations, metaphor, analogy, their expression in totemic or magical belief, and correlates in dream and psychopathology—adaptations to conditions, inner and outer, that are meaningful to the individual even if, at times, they seem to others to be irrational. The importance of the non-rational is that, in varying manifestations, it is a recurring mode of thought that is active beneath the surface of everyday behavior and conscious reason. A truth is a belief that is true to the believer, but belief is not truth. Indeed, science helps to muzzle false belief in adhering to evidence and demonstration, but truth is not a measure or surrogate for belief, it is a goal and a process, with the aim to dislodge other beliefs that have been shaped by personal desire and experience, the innate derivatives of a primitive mentality or habitual patterns of thought instilled by coincidence and enforced by indoctrination.

---

2   An example in Lucas (1970) is the statement that Freud's theories are "ridiculous" without an obligation to justify this remark.

Terminological Confusions

Confusion about certainty — of belief, knowledge — is partly due to the loose usage of various terms, such as, "I know, believe, think, feel, suppose" and so on, all of which, on occasion, especially in popular culture, are used interchangeably, so that some clarification, or admission of irreconcilable confusion, or a search for another vocabulary is in order before going more deeply into the matter. Statements such as "I believe" and "I know" are used in similar and contrasting ways. We can say "I believe" to express uncertainty, as in "I believe there is no water on Mars, but I am not absolutely certain", for it is conceivable that water on Mars will one day be discovered, or we can say "I believe in god, the devil, an after-life", in which belief is strong without a hint of truth or possibility of disconfirmation.

The use of "I know" is similar, in that to know something can, as in science, be provisional, e.g. "I know that all horses have a tail but I could be convinced otherwise." Unlike conviction in belief, which often resists argument and evidence, the certainty in knowing is time-conditional. Certainty could claim "I know it is raining today", while belief and conviction tend to transcend locality. To say "I know" is often comparable to say "I believe". In general, to have a belief or know a truth is to possess some degree of certainty that the belief is true. In this way, *truth* tends to divide up knowledge and belief. Wittgenstein (1969ed, [243]) wrote, "'I know' relates to a possibility of demonstrating the truth." However, one can know or believe various things without knowing if they are true. Truth is presumed indispensable to certainty, but many people are certain about a truth of which others are unsure (e.g. eating fish is healthy), and there is doubt for, or resistance to, things that have been shown to be true (e.g. aiding the needy promotes dependence). At one extreme is a "truth" that lies in absolute faith, often religious. At the other extreme is a truth presumed to be absolute and eternal in some modes of knowing, e.g. mathematics, where one can say the truth is timeless. Here, to know means to know with absolute certainty, and to believe means to believe with absolute conviction.

The addition of "I know" or "I believe" to a statement such as "I see, I feel" is a way of admitting doubt or reinforcing certainty, but there is confusion on this point. One would not want to say "I know I have a pain" instead of "I have a pain", even if this is an acceptable statement, since "I know" leaves open the possibility of error, and it seems unlikely one could be mistaken about having a pain. Yet there are conditions in which having a pain is questionable. A pre-frontal lobotomy case might say, "I have a severe pain but it does not bother me." This is not stoicism. The pain has lost its noxious quality. What is a pain that is not painful? The person could justly wonder if he is or is not in pain. I have dreamt I was in pain, or awoke briefly in pain during anesthesia and was confused as to whether this was "real" or imagined pain. This does not differ significantly from saying, "I know I see a tree" instead of "I see a tree." The doubt in the latter is credible since we are accustomed to hallucinations, misperceptions and illusions, so one could say, for example, "I know I see a tree in the distance (and not a mirage or a rock formation)." Vision and pain are both perceptions, one nociceptive, referred to the body, the other exteroceptive, referred to the world. We are generally more confident of bodily states than states of the world.

Take the example, "I know which mushrooms are edible." This knowledge consists in remembering a fact that may have come from direct experience, e.g. instruction, reading, getting ill from eating a mushroom. One could say—in a manner that is odd but not implausible—"I believe I know this mushroom is edible", where belief stands for uncertain knowledge. It is equally plausible to say "I know that I believe this mushroom is edible", where one is certain of having a belief even if the belief itself is uncertain. Certainty in belief is tied to fact at the objective pole and to faith at the subjective pole, while certainty in knowledge is tied only to truth. Fact is presumed to be equivalent to truth, or to a best approximation, but it is also an endpoint in valuation. The substantial overlap in these mental activities, with a mixing of terms and the phenomena to which they refer is a challenge to philosophical discourse in that it reflects the coherence of the system through which acts, feelings and objects develop.

A psychology of knowledge and belief from the standpoint of microgenetic theory is not particularly concerned with truth-judgments that depend on the relation of appearance to reality or the correspondence of statements about facts with the grounds of facts in reality. This raises the question of fact in relation to value and introduces an epistemic problem discussed at length in other works. From an internalist perspective, the boundaries of truth, knowledge, fact, value and belief are porous and continuous, with each term designating a state that is heavily interdependent.

Certainty and Reason

Certainty has a relation to truth; conviction has a relation to belief. Truth-judgments attempt to be mind-independent. In the *Principia*, Whitehead and Russell went to great lengths, some 300 pages, to demonstrate that one plus one is two. If one is a unity, two unities are not a sum but a novel oneness, as implied in Whitehead's writings.[3] The mind-independence of mathematical logic is achieved at the cost of its subjective ground.[4] Ambiguities at the quantum level point to the subjectivity of mathematical logic. I think it likely that philosophy of mind and theory in physics are guided by common laws or principles.

The truth of a belief that is validated by science and/or consensus plays a modest role in the *feeling* of certainty. The certainty of a truth, to the extent there is certain truth, is no guarantor of belief. To the rational mind, certainty is reliable knowledge that consists in facts and experiences concordant with demonstration, argument or observation. Coherence comes of the recurrence of similar occasions, from regularity and uniformity and the relative predictability of outcomes. Experience is assimilated to implicit knowledge. At each moment, mind and world are inexact replications. The inner and outer portions of the mental state are under constant assault by novelty. Similarity of occasion — coherence, assimilation — is one aspect of certainty, e.g. that the sun will rise each morning. In contrast, dissimilarity, changeability, e.g. the weather, is the basis of uncertainty, at least in the form of predictability.

Those accounts of certainty that rest on truth do not penetrate deeply into belief, which is narrowly confined to a system of knowledge in which linguistic or mathematical propositions are embedded, and far removed from life-experience. Certainty is the enemy of novelty. In a premise, the immediately given, e.g. a syllogism or formula, is all there is to work with. Whitehead (1933, p. 313) alludes to the absence of the creative, in writing, "it is more important that a proposition be interesting than that it be true."[5] Logic is an artificial isolate in the dynamic of thought that has little to do with actual thinking. As is well-known, Einstein said his thought consisted mainly of visual images. The cognitive embedding of truth-judgments is what the philosopher Crispin Wright referred to as an "information-dependent warrant for a particular proposition".

---

[3] For example, Whitehead's (1929) well-known comment that "the many become one and are increased by one."

[4] Mathematical propositions are usually taken as exempt from doubt, i.e. incontrovertible, but Wittgenstein (1969ed, [651]) writes, "...one cannot contrast *mathematical* certainty with the relative uncertainty of empirical propositions. For the mathematical proposition has been obtained by a series of actions that are in no way different from the actions of the rest of our lives, and are in the same degree liable to forgetfulness, oversight and illusion."

[5] According to Lucas (1995) this passage, one of his most "widely quoted scandalous", points to his later skepticism, as with that of Wittgenstein, as to the certainty of truth.

A more expansive proposition, e.g. a horse is an animal, entails a wider set of relations that include implicit concepts of horse, animal, category and member, of being a category and belonging to one, boundary conditions, e.g. prototypical (characteristic) and atypical features, as well as existence, tense and time, e.g. small horse-like ancestors, past and present horses, imaginary horses (centaurs, unicorns) and so on. To know a horse is an animal is to have knowledge of the concept of a horse and its relation to wider experience, as well as an intuition of self as agent, the reality of an external world that includes horses and other animals, and unconscious presuppositions that concern the nature of things, subject/object relations and valuations. "Information-dependent" to say the least!

The syllogism is a variant of logical reasoning but here, too, statements are nested in conceptual and experiential knowledge. Socrates is (was) a man, implies knowledge of the existence of Socrates as opposed to Poseidon or the tooth fairy, i.e. a distinction of actual (past or present) existences from mythic or imaginary ones. "Is a man" implies adult and masculine though it refers to human, while mortality depends on religious belief and the sense in which life ends, e.g. the "immortal" discourses of Socrates, genetic continuance in children, cell-lines, cloning, the possibility of arresting the aging process, even frozen corpses. Other systems of logic challenge the convention, such as the tetralemmas and five-pronged logic of Buddhist philosophy. Again, the point is that a great deal has to be excluded for the syllogism to obtain, much less to be true.

The untapped potential in a syllogism is evident in paralogic, dream, magical and psychotic thought, which takes the form:

Christ has a beard
I have a beard
I am Christ

Tigers are fierce
Akiba is fierce
Akiba is a tiger (or has tiger spirit)

In logic, the predicate is a property of the subject with an inference based on this relation. In paralogic, the subject is a property of the predicate, such that shared attributes are the basis on which different subjects are identified (von Domarus, 1944; Brown, 2005). It is well to keep in mind the animistic roots (Lévy-Bruhl, 1935/1983) of syllogistic or subject-predicate reasoning (Whitehead, 1933), for reason develops

—not just historically but momentarily—out of a non-rational ground that is excised from its skeletal remainder.[6]

The dilemma is that a truth established by reason and taken as informationally encapsulated, i.e. mind-independent, says less than assumed about mind (or world), since reason and the residue of knowledge behind it are thoroughly infiltrated by need, experience, belief and valuation. The salient fact is that the isolation of propositions from the very context that needs explication is a ground and justification for skepticism. The ascertainment of truth in formal logic turns an assumption into an assertion and, by separating content from psyche, weakens the explanatory power—the full truth—of its conclusion. A sign of this limitation is that truth achieved by reason does not mandate conviction in the same way, or to the same degree, as truth motivated by belief. This agrees with the argument that truth develops on a bedrock of *dispositions and presuppositions* or trends in tacit knowledge (Polanyi, 1958). For Wittgenstein (1969), truth is not simply the outcome of a judgment but requires the *acceptance* of a statement as true. Acceptance is recognition of the basis of theory in presupposition (Collingwood, 1940). If the felt context of a statement cannot be examined, a deeper and more incisive truth cannot be eliminated.

Knowledge and Belief

Certainty lies at the intersection of belief and knowledge. Knowledge is, broadly, what we know, not what we believe, since much of knowledge has little to do with truth, whether knowledge of a vocabulary, knowing what one does not know or needs to know. Beliefs may or may not be supported by knowledge. A belief can refer to the uncertainty of what is known, when "I believe" is weaker than "I know", as in "I believe that… she loves me, it will rain tomorrow, pigs are intelligent", though I don't actually *know* these things, or belief can justify certainty (conviction) regardless of the truth, as in faith, superstition, strong opinion, fanaticism, delusion, etc. In knowledge, there is a distinction between knowing "what" and knowing "how", or fact-knowledge and skill.[7] One knows a language well or poorly, as one knows how to drive a car. *Knowing how* to speak or how to drive is not a matter of belief, though *fact-knowledge*, e.g. that leopards have spots,

---

[6] In philosophy, the irrational has been discussed, e.g. Mele (1987), but finds no home in the genesis of thought.

[7] Whitehead (1929) discusses two modes of reason, that of the gods (Plato) and that of the foxes (Ulysses). The practical reason is skill-based, closer to action and the animal inheritance, while speculative reason is conscious, distinct from action and a "higher" mode of knowing.

can approximate belief. One could say I believe leopards have spots instead of saying I know this, where knowing implies greater certitude than believing, as in: I believe some leopards do not have spots but I do not know this "for sure", or I believe IQ is inherited but the matter is yet undecided. Belief approaches knowledge when an erosion of coherence weakens conviction. In the main, however, the glue of conviction is the inculcation of value and the fusion of disparate concepts by relations of predication that align topics on the basis of shared properties. This is why, in part, conviction is not readily altered by fact, and why knowledge can lack conviction in spite of a preponderance of evidence.

Some beliefs appear to be unshakeable, e.g. I believe I am alive, or "rock bottom" beliefs such as that expressed in Moore's statement, this is my right hand, but there are cases in which people believe they are dead (the Cotard delusion), amputees feel and "move" a phantom limb, and autotopagnosics (Brown, 1988), asked to show their hand, may say "it is over there on the wall." Conviction depends as much on the collusion of the senses as on the coherence of knowledge. Thus, an auditory hallucination, e.g. a voice, may seem imaginary until it involves the visual modality, e.g. a face, at which time the individual believes the image is real (Hecaen and Ropaert, 1959). In dream as in wakefulness, perceptual conspiracy reinforces the realness of singular modalities.

In maturation, relational systems of knowledge and belief accumulate a vast array of implicit and explicit facts and observations to which new experience is assimilated. An observation that is dissonant with this system, e.g. a three-headed dog, induces a state of wonder or perplexity, or one of tension that is resolved when the observation is explicable as a genetic mutation, an aberration, a fraud or illusion.[8] In such a situation, one might say, "I can't believe my eyes", so either an explanation is required or the category is enlarged to include dogs with three heads. The individual with a flexible knowledge base — someone we say is open-minded — can entertain competing or discordant beliefs, but to a limit. For example, we know from scientific reports or sideshows that an individual can have two heads and multiple limbs, but that does not mean we are prepared to accept as other than a fable or metaphor the discourse on love by Aristophanes in the Symposium on

---

[8] Recently in India, a child born with multiple limbs was believed to be the incarnation of the god Shiva, and many gathered to pay respects. The departure from ordinary experience was interpreted as a miracle. Similar occurrences have been reported in other uneducated or impressionable peoples, such as the reverence for an albino in dark-skinned cultures.

the splitting of an original being in two parts that seek their complementary halves.

For the ordinary person, an opinion that is based on incomplete or faulty knowledge can be stronger than a truth based on science or logic. Such beliefs or opinions, for example, that "health foods" are healthy or that extraterrestrials have visited earth are less bound to the knowledge base than proven facts but they are accompanied by a feeling of certainty that is often greater than that for true knowledge. Conviction is more powerful for questionable or fictitious beliefs or misconstrued observations than for facts established by reason or science. Conviction resembles faith in the satisfaction of unconscious need to which facts do not appeal. A truth established by fact attempts to be extrinsic to need and can accompany a certainty that endures until the belief is disconfirmed or replaced by new facts. With conviction, facts are twisted or ignored to protect the belief; in certainty, it is the reverse. Conviction is an outcome of the personal need that is the bane of impersonal reason and not a matter of naiveté or gullibility.

Scientific truth, which is not apprehended as absolute but as plausible, does not engender the conviction promulgated by implicit value, magical belief or personal experience. While scientific facts may be defended with vigor and precision, and have a greater license on truth than popular opinion, they are conceded to be open to disconfirmation; indeed, the goal of science is to make yesterday's facts obsolete. As Gödel demonstrated, one can doubt a logical truth that appears unassailable, while in a similar vein, and in the spirit of Hegel, Niels Bohr wrote, "the opposite of a trivial truth is false (but) the opposite of a great truth is also true."

Science rests on a set of shared assumptions, a guiding paradigm or *Zeitgeist*, of which ordinary people have little knowledge and less interest, at least not in an organic and deeply felt way. Even if scientific research has a significant impact on people's lives, e.g. progress in medicine, technology, the parochial values of science do not penetrate the core needs of personality.[9] There are many ways people are affected by scientific truth but, not withstanding greater health and longevity, none bear on personal fulfillment and happiness. The publicity of rational truth dilutes the privacy of conviction. A deep-seated belief differs from dependence on knowledge and the need for validation prior to conviction. The former is inculcated largely in childhood; the

---

[9] See Dickinson (1937) for an elegant discussion of this topic, including the dangers that ensue when religious beliefs are eclipsed by scientific authority.

latter is reached by mature deliberation. Reason seeks appraisal by others in a quest for impersonality, assigning truth to judgment.

One cannot readily go from a reasoned argument to a decided fact or from a consensual fact to an uncompromising belief. A rational argument that is compelling and persuasive rarely leads to a conviction that, in intensity, compares with untutored belief. For that matter, even scientific authority is trumped by the certainty extended to the "prophetic truths" of dream. The stronger the belief, the less truth matters. Indeed, the trick of a leader is not to convince followers of a truth, but with slogans and appeals, often to the basest of values, to arouse emotion and conviction in a cause. A person will sacrifice his life for a belief that, to reason, is false or inane, but will barely tolerate an inconvenience for a true fact. To believe in something you think is true is not to believe in something you know is true, but even true knowledge, to be *felt* as true, must cohere with the manifold of the inner life.

A rational knowledge base is not an impervious defense to irrationality. Reason is the standard of thought but not an arbiter of conviction. Even in sophisticated thought, the rational and irrational are deeply and invisibly commingled. Poincaré famously wrote that he had to laboriously work out a mathematical proof, the truth of which was given by intuition. Many examples of creative thought in dream and transitional states can be found in Koestler (1964). William James, writing on Fechner, argued that philosophy was more a matter of passionate vision than logic, logic only finding reasons for the vision afterwards. I have written at length (Brown, 2005) that reason is not the instigator of action but the *justification* for acts of unconscious origin. The relevant distinction is not knowledge and belief but unconscious predisposition and conscious reflection. Affectively-charged beliefs and tacit knowledge are derived to concepts that evoke conscious judgment.

In sum, truth with certainty and belief with conviction are decided by experience, but subject to patterns of thought, valuation and relations of predication, i.e. the whole-part relations that stand behind and specify judgments. When the subjectivity of a judgment is prominent, there is conviction; when objectivity is prominent, there is greater or lesser certainty. Truth, which hangs in the balance, lies in the conformance of appearance to reality, or the adaptation of mind to world. An idea or a statement is true if it conforms to facts that are consensual, thus external. A system of factual knowledge is built on the adaptive resolution of instinct. Dewey put it well when he wrote that facts are irreducible values.

Truth and Knowledge: A Speculation

The subjective edifice of mind is stratified into layers of affect, imagination (dream, fantasy), thought and perception, with all layers — successive approximations to a model of the world — participating in every act of thought. At each level, a world is created, but truth is conformance across layers. For example, animism and dream present a vision that is more or less real as it conforms to waking perception. This is not correspondence but the gradual adaptation of inner to outer, *a transition or becoming from image to object, from idea to fact or from fantasy and the creative imagination to action and reason.*

A pragmatic approach to truth (James, 1907) has the limitation that the perceptible world (however accurate the model) is still a representation of the unknown. The truth in conformance — a model and the reality it approximates — can only be adduced by adaptive success, and thus is inexact, even if it is the only truth we can know. Conformance owes to a "veto" of the maladaptive so only that which shapes the model to the world survives. Conviction accepts a conformance of belief to world that reason might disown. Certainty aims at a conformance of segments in the mind/brain state distal to the origins of belief (Fig. 8.1). The adaptive values of logic or science that underlie certainty are testing grounds of truth, not conviction. In conviction, personal need is foundational, while certainty, which is based in (perceived) fact, depends more on consensus. Conviction is more like confidence; certainty more like acceptance.

The veridical is the truth in perception; the context left behind is the ground of art. Perception resolves image with object, art resolves need with satisfaction. The momentary context of perceptual experience is, more deeply, a pathway to the creative. Art is a contact that adjudicates belief. Perception resolves belief with reality; art takes reality inward and resolves it with belief. The immediacy of perception is the hoped-for in art, but the interior of perception is a search for outer constituents while that in art is inner signification. The allusive or revelatory nature of art is, in perception, just waiting to be uncovered.[10]

Knowledge (experience) is not mere information; every "data-point" rests on, and includes, a vast multitude of potential configurations that incorporate what we remember and use, what is implicit but forgotten, what arouses or is recalled, what inspires, what informs and what is irretrievable. On the basis of techniques such as hypnosis, some

---

[10] The contextual background of statements contributes to their uncertainty. Wittgenstein (1969ed, [623]) wrote, "...in such a case I always feel like saying (although it is wrong): 'I know that — so far as one can know such a thing.' That is incorrect, but something right is hidden behind it."

theorists (McCulloch, 1965) have speculated that all experience is preserved as a kind of *Lebensfilm* or "movie" reel.

While knowledge is parsed by belief to what is valued, belief serves knowledge when it passes into fact. The mapping of belief to world is inessential to conviction, for belief enjoys conviction without the obligations of truth. Knowledge can motivate perfunctory belief in true facts but not a conviction that agrees with unconscious presuppositions. Primitive belief such as fear of spirits or worship of totems is not just a pre-scientific explanation of nature but a coping strategy of early cognition. Knowledge and experience arise out of the instinctual repertoire; they are illuminated by value and categorized by belief (Fig. 8.1). In sum, *the function or role of value is to give valence to knowledge; the function or role of knowledge is to parse belief into what is adaptive; and the function or role of belief is to give conviction to what is known. The coincidence of belief with knowledge and the reinforcement by valuation and truth are measures of successful adaptation.*

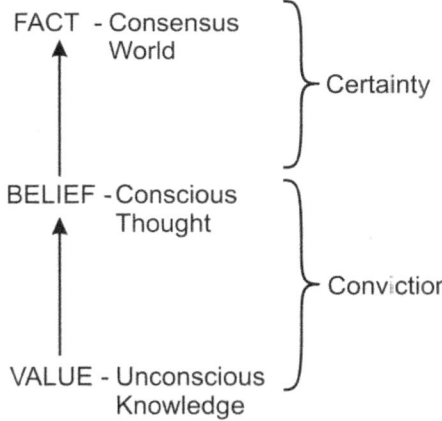

Fig. 8.1: Conviction arises as belief penetrated by value (feeling). Certainty arises as belief shaped by fact. The distinction of conviction and certainty reflects the degree to which belief is driven by need (value, feeling). In the former, value is pre-eminent; in the latter, the distillation of belief in thought gives choice. This reduces the impact of feeling, allows belief to adapt to the world and partition to fact. Belief is the categorical shape of objects; value is the processual dynamic of belief. In brief, certainty is value distilled by belief and validated by adaptation and/or consensus, while conviction is belief motivated by value in the satisfaction of need.

Limits of Belief

Certainty refers to the belief that knowledge is factual or true, while conviction is strong belief in certainty, or certainty that no longer entertains doubt. Unlike conviction, which can be independent of fact, certainty is often the product of demonstration, observation or trial and error. A main characteristic of conviction is the exclusion of choice. Strong belief does not tolerate options or uncertainties. Fixed opinion is like conviction in that options are rejected or ignored. This is not necessarily undesirable. The adage, which I believe owes to H.G. Wells, that one should keep an open mind until it is made up, then close it, guarantees acceptable decisions based on personal values that are applicable to a variety of situations, so that uncertainty does not plague every choice.[11]

Strong belief can be inimical to reason but even irrational beliefs can lead, at times, to rational outcomes, e.g. Kissinger's remark that even paranoids have real enemies. The beliefs of most psychotics are false, at times obviously so when they involve little green aliens, but at other times plausible. Yet, a schizophrenic for whom the world is unreal, who confuses thoughts with perceptions, who is uncertain of the continuance, agency and autonomy of a self, may well be closer to the truth of statements about mind and world than the common-sense beliefs of the ordinary person for whom such confusions do not arise. Indeed, much of the bizarre content of psychotic thought comes from a mix of image and object, dream and perception, with an emphasis on phenomena associated with dream, not those of rational cognition. It is fair to say the false beliefs to which the psychotic gives expression derive from phases that in most people are ordinarily submerged.[12]

When irrationality generalizes in scope, in content and to multiple occasions, truth may seem accidental, not the outcome of reason but serendipitous and buried in fantasy and excess. Yet *at a given moment*, what is the difference in the mental state of a rational person with a true belief, say, that he is persecuted by the IRS, and a paranoid with a similar belief that happens to be true? One belief is based on real events, the other on the imagination, but in both instances the belief points to real facts. However, in the paranoid the facts, though true, are generated by the belief, while in someone who is rational, the belief is presumed to follow the facts. It is not only physics but in everyday life that Einstein's remark applies that "it is the theory which decides what we can observe [the facts]."

If the factual basis of a belief is the warrant of its truth, the belief, e.g. persecution by the IRS, is true independent of how the facts

---

[11] The question, "Is it true?" differs from, does A explain the facts better than B especially when a more persuasive option, C, is not mentioned or imagined.

[12] Wittgenstein agrees with Moore that some beliefs are so obvious only a madman could doubt them, but in delusion the issue is less the presence of doubt for true beliefs than the absence of doubt for false ones.

originated. If the outcome of the reasoning that supports the belief is correct, albeit logical in one, irrational in the other, and if both beliefs are true, what is the deciding factor? If the psychological underpinnings of schizophrenic belief distance it from rational belief, why should psychological underpinnings be excluded from logical statements? Ideally, reason requires choice—even doubt or uncertainty—but thought need not be deliberative. Imagination is often not rational yet experienced with certainty. One could say, the whole basis for assuming a rational decision is a sifting through arguments or evidence to settle on the most logical or reasonable choice. But, as stated, this approach, which is the way of scientific thought, at best provides an approximation, not a truth, depending on the coherence of the facts selected and a lack of more persuasive explanations.

In the paranoid, a persecutory belief, even on the occasion when it is true, can perhaps be said to be false if there is a poverty of evidence or a lack of specificity to a given instance, i.e. context, related beliefs that are patently false. If a person believes the IRS, FBI, CIA and KGB are after him, the lack of specificity alone is enough to nullify the belief. If one element of the belief is false, it tends to taint all others. If I believe that tomorrow there will be sun, rain, clouds or snow, the belief, though containing a true statement, cannot be said to be a true belief. In other words, beliefs have to be coherent, in this instance as to time, space and setting. It is meaningless to say, "I believe it will rain", without implying or denoting a place and time when this will occur, for it is always raining somewhere, sometime; as it is meaningless, unless a time and/or place are given, to say, "I believe I will die", when this is a definite eventuality.

The example of paranoid thinking raises a further point, namely that novel ideas defy the ingrained beliefs of others and, in rational people, their own. To challenge the accepted beliefs of others could be a sign of genius or lunacy depending on whether the novel belief is true, which may be uncertain, as in new science, or when the competing belief is irrational, which may be in doubt. What do we make of a rational person who arrives at a true belief by reasoning that is incomplete, misguided or not fully rational, such as Kekulé's dream of snakes that inspired the carbon ring, or conscious reasoning that is slightly off-kilter, such as the evolutionary theory of Erasmus Darwin, or the pre-Freud embellishments by von Hartmann on the unconscious? What of the converse, a person who arrives at a false belief by logic that, at the time, appeared impeccable, such as the belief in the Piltdown hoax or the justification by many scholars of disbelief in the significance of the *Taung* skull? Reason is a guide to the truth of a belief, but reason and truth, or the aim to truth, do not always coincide.

Entire systems of belief, from Marxism to Capitalism, from psychoanalysis to cognitive science, fail or succeed not because they are true or false but on their value to individuals and satisfaction of need that determine the outcome of a struggle with competing paradigms.

Truth and Belief

Wittgenstein mixes the problem of belief with the knowledge of reality. One could say, all beliefs about reality are acts of faith, with all knowledge of reality an approximation based on such acts. But he does not want to say that certainty is an approximation to a constructed point. Rather, doubt gradually loses its sense (1969ed, [58]). If doubt loses sense because the reasons for doubting are cleared away, and this increases certainty, it still leaves certainty as an approximation. When certainty of knowledge will not allow the disclaimer "but I could be mistaken", it approaches belief, not as provisional knowledge but as opinion that is insensitive to reason. This can lead to belief that is oblivious to truth, from which it is a short step to delusion. However, the certainty that results from an *absence* of doubt may still not entail conviction. One can be certain of a truth but not believe it deeply. I am certain Mongolia is cold in winter but my conviction is tempered by indifference, global warming and lack of personal experience. Only when certainty is punctuated by personal valuation does belief — true or false — become unassailable.

Belief takes on significance, as *individual* belief, when place and period are part of the relevant event. On the other hand, a powerful belief gains credulity when it is time-independent, i.e. a timeless truth, such as those of mathematics. To say, I believe it has rained in the past is true but trivial, while to say, I believe it rained in New York yesterday, is meaningless, since either it did or did not rain, so the belief is merely an admission of uncertainty or ignorance. To say, I believe 1 + 1 = 2 can be a confession of ignorance or the insight of a great mathematician. A true belief is a statement of fact, though statements of fact can be framed as propositions requiring truth-judgments, after which one can say I believe the statement is true. On the one hand, certain belief is knowledge that is true or false, yet for the individual, beliefs may possess a certainty unsupported by fact.

A belief depends as well on presuppositions that, if true, give credence to the belief and its subsidiary content or, if false, arouse skepticism in others. To a great extent such doubts are valuations supported by consensual assumptions. Swedenborg (1758/1940) gave a detailed and tightly argued account of the architecture of heaven that was supported, he wrote in the introduction, by a conversation with an angel. This example, though dated, seems extreme, but it is similar to modern pronouncements about the physical world that do not consider the illusory or mind-dependent nature of perception, or the widely accepted belief that the brain operates like a computer. Such statements are not statements of fact, as assumed, but speculations that depend on presuppositions—false in my opinion—that a real world is directly accessible to perception or that the brain is, or is like, a computer. Such examples teach us that the unstated in a belief is the ground of its refutation.[13]

Doubt, not error, is the contrast with certainty. Wittgenstein asked if one needs grounds for doubt and went on to suggest that doubts form a system comparable to beliefs. Can doubts form a system the grounds of which consist of presuppositions and the knowledge that supports and flows from them? Is doubt nested in a more or less coherent set of relations, like beliefs? Is doubt like this? It seems the grounds of doubting entail conflict with the belief system to which doubt must—but is unable to—assimilate. This does not imply a system of doubt, yet it is conceivable that one could have a widespread skepticism that forms what might be called a negative belief system.

Value, Fact and Knowledge

The dependence of fact on knowledge, and the relational quality of that dependency, means that every fact depends on other facts, true, false or uncertain. The belief that a fact is true, or a disputation of its truth, depends not only on this relation but its value to the individual. Ultimately, valuation decides what facts are important. This is why reason is most efficacious and belief least assertive in matters about which we care little. Scientists work with technical problems that for the most part are distant from life-concerns. The passion in some scientists for a particular (usually their own) belief is more likely to represent a drive to power, ambition and a view of skeptics as enemies, rather than a forceful belief in the truth of the asserted facts.[14]

If we take the opposite path and try to imagine an impersonal reason devoid of character, bias, loyalties, values or interests, i.e. uncoupled from personal history, like the self *an sich* of Kant, on what basis could any decision be made. The search for truth in science and in life must contend with the history and values of those who are search-

---

[13] Perhaps what Wittgenstein (1969ed, [253]) had in mind in writing, "At the foundation of well-founded beliefs lies belief that is not founded."

[14] To the point, the witty remark of the astronomer Harlow Shapley, "A hypothesis or theory is clear, decisive, and positive but it is believed by none but the man who created it. Experimental findings [facts] on the other hand are messy, inexact things which are believed by everyone except the man who did the work."

ing, which are not readily extracted from a consensus of like-minded authorities. Far from being independent events in the physical world, facts are the final phase of personal and interpersonal valuation. On the other hand, if every act of thought begins with instinct, for which I have provided a framework (Brown, 2012) and as Wittgenstein and others have argued, what are the grounds of rational decision? The effects of instinct are not arrested at the first inklings of thought, but shape, focus, limit and suffuse the concepts and feelings of the primitive drive categories that condition thought at every phase in its formulation.

The pretext of impersonality is essential to truth and fact. We think that facts are neutral and take on direction or importance in relation to judgment, but this is not exactly true. Every fact and experience is influenced by value, in perception, interpretation and retention. To remember an event implies its valuation; minimally as interest, like or dislike. Otherwise, the event—the fact, the observation—would not have been remembered, perhaps not even perceived. Value is a mode of feeling that arises in drive, individuates to desire and other affects, and goes out as worth into the world of objects. To state a fact is to uncover the value that imbues it with significance. Value determines the selection of an experience, a memory or feeling, and whether or not it becomes conscious. To argue the truth of a fact is to exhibit the valuation invested in it. Value is *not attached* to mental or external contents, nor does it accentuate other feelings; it is their affective tonality at serial phases in the subjective aim. Value accompanies the formative process from the inception of the mental state to its final execution and at each phase fully inhabits the idea, fact or object that actualizes. In the arc of microgenetic process, value is the decisive factor in the actualization of belief.

A theory of belief and certainty cannot ignore the central role of value in the determination of truth. Value gives import to facts that constitute the knowledge base. The fact becomes true when its value is supported by other facts and not disputed by contrary ones. The valuation that gives conviction recalls Nietzsche's comment that truths are irrefutable errors, but a lack of refutation is insufficient to turn a value into a fact. The inability to find water on Mars, or to refute a statement such as "there is no water on Mars", is not enough to make the statement true. As evidence accumulates in favor of the statement, and no evidence is raised in opposition, the statement will eventually become true, even if it was true all along independent of confirmation.

The notion that truth-finding is an effort to reach an unimpeachable truth such as $1 + 1 = 2$ that is forever and tenselessly true, implies that at some point, when the process of inquiry is suspended, in life, in science and in philosophy, truth becomes settled, i.e. conforms to a truth that is timeless, that is, a truth that was, is and will always be true, but at what point does this occur? Most often this is, as Wittgenstein put it, a matter of acceptance, not certain knowledge. Acceptance entails a valuation (evaluation) of the facts at hand. This is related to confidence, which has a quantitative feel, as if it were a degree of acceptance. Wittgenstein uses the term *sureness* for the immediate acceptance of fact. We speak of the *weight* or *preponderance* of evidence, but the weight given to facts depends on the value alloted to them. In brief, *facts without evidence are hypotheses or beliefs that can be shown to be true or false by other facts. When the valuation in a fact develops to a certainty that compliments the evidence in its favor, it is a belief. Certainty is the valuation of fact as true knowledge; conviction is the valuation of true or false knowledge as fact.*

The distinction of implicit and explicit is central to all forms of mental activity; memory, thought, language and so on, with a widespread consensus on the distinction. It is well-studied in memory research and is embodied in language in the relation of competence to performance or grammatical knowledge to its use. For some, the distinction points to separate stores, components or anatomical systems, while microgenetic theory postulates transitional phases. In all modes of cognition, what is implicit refers primarily to tacit knowledge or memory, while the explicit refers to the path taken when the implicit becomes conscious. Knowledge is to thought, action or remembrance as skill (unconscious thought or memory) is to performance (consciousness of implementation). The use to which knowledge is put determines the mental category to which it belongs. Implicit knowledge that develops to replication is memory or recall. If novelty predominates, it is thought or imagination. Only a fraction of implicit knowledge becomes conscious.

For example, as I am writing these passages the memory of what I have already written, or thought about, the thoughts of others on this topic, personal experience, knowledge of language and the feeling invested in thought, are all implicit or tacitly present in the background of any idea that develops to consciousness. A portion of this background can become conscious, but when this occurs it is experienced as a search or an appeal, not as a personal or novel thought. To be conscious of the known is to remember; to have a memory that is in the service of conscious thought is thinking.

The categorical nature of knowledge—indeed, of all mental contents—accounts partly for the elusiveness of truth, since fact is the final-most point in category-resolution. A fact, or the category behind it, expands by arousing context that subsumes virtual or potential facts that may reinforce or destabilize its truth. The inclusiveness of a category can account for doubt and uncertainty; its exclusiveness, for the resistance of belief to contradiction. The category generates data as an antidote to refutation. In brief, *when a decision as to the use or trustworthiness of knowledge is required, in so far as it agrees with estimations of reality, knowledge can be said to be true or false. Knowledge becomes value by virtue of its selection and the use to which it is put, and it becomes belief in relation to the conviction in its truth.*

Chapter Nine
Psychology of Free Will

*If we do not know what it is,
begin with what it is not.*

When all other mental phenomena have, arguably, been reduced to brain events, free will is the last retreat of the dualist and the nexus of the problem of dual-aspect theory or mind/brain identity. In a previous work on this topic (Brown, 1996) I cited my late friend, Detlef Linke, who put the issue in a nutshell when he asked why I was so determined to write on free will! Since then, in searching for some path to agency independent of brain or causal ancestry, my focus had been on the virtual present that hovers over the "instantaneity" of physical passage. The question was whether the succession of phases in a series of overlapping mind/brain states might be influenced by the self in a virtual duration that arises in the present state but spans the series of states it envelops. The before/after of causation is suspended in a present created across the succession of states and over successive phases within a state. The central point is that the serial order essential to causation is generated out of a state in which the before/after is simultaneous.

Put otherwise, if as I believe, (1) the present is derived from the disparity of the endpoint of a state with the revival of a previous one, and (2) the self is a precursor of all consequent events within the duration and (3) the duration subsumes a series of epochs, given the simultaneity of each state until completion, the unfolding to serial order that is apprehended as change, and the recurrence of the self at the liminal floor, could the series that is embedded in the present, which represents the overlap of states within the present, be controlled by the self before the state actualizes? The feeling of agency would be for a relatively stable self in relation to changing mental contents that are nonetheless simultaneous prior to actualization.

I have since retreated from this account, less for lack of evidence than theoretical uncertainty, but it still seems a valuable line of inquiry. The approach did raise, as do all others, fundamental questions on mind and brain that impact the notion of agency. These include: can all mental phenomena be reduced to causal brain activity; do consciousness and free will appear (?emerge) at some point in the evolution of complexity, what are the parameters of freedom, what is the nature of will; and what is the distinction of choice and decision from decision and action? One has to address the nature of the self, its origin, locus, causal role and dependence on personal history and circumstance; the ancestry and range of options; the relation of self to choice, decision and action; the definition of free will and agency; the compatibility or detachment of will and causal brain process; and, critically, whether brain activity is invariably causal.

The mental life is replete with illusions (Vaihinger, 1924/1965) so tightly adapted to the exigencies of survival they might as well be tokens of reality. The belief in mind-independence of the perceptual world is one. Is a belief in the brain-independence of a self another? Whether the freedom in free will is illusory or real, how can we account for its occurrence from the standpoint of psyche and brain? If free will is an illusion, the obligation is to explain its source as one of the most powerful illusions we have. If the self is illusory, like the *specious present*, so are mental contents that fall within its duration and on which the self is felt to act. The idea of mental causation from one illusion to another would seem to be nonsensical. The self cannot be dismissed as an illusion, while an effect of mind or consciousness on action is retained.

Central to the discussion are the parameters of choice, the latitude of options that could satisfy conditions for free will and the micro-temporal phases that lead to decision and action. Could a self as a causal outcome freely decide and cause an action, or is this determined by unconscious antecedents or concomitant brain activity within and across mental states? Some arguments are set before the reader that as a whole lend support to the possibility of free will. The problem remains dicey, but so do arguments for determinism that assume universal causation in spite of an absence of — even disdain for — psychological data and the lack, except for microgenetic theory, of a model of the mind/brain state.

What is Free Will?

What does it means to have free will?[1] Kant (1780/81) noted the contradiction in an impersonal self that is motivated by pure reason unguided

---

[1] Freedom pertains to the limits on action once decision is made, while free will entails making a decision or consciousness of choice. The feeling of volition is distinct from agency. Volition is the feeling of choice and agency.

by bias or personal experience. Such a self would have no qualities to effect decision, no loyalties, preferences or affective interests. Perhaps machine intelligence could meet these criteria, as in the selection (computation) of a move in chess, but most decisions are neither impersonal nor rational, such as to go to a concert or ballet, see this film or that, eat one food or another. These are not moral or logical dilemmas but they involve agency no less than a moral judgment. Indeed, a simple impersonal decision such as to lift a finger without evident aim or provocation is a paradigmatic example.

The self is permeated by will as a derivation of instinctual drive. Will is the forward impulse of drive, modulated by the self and specified to partial manifestations such as desire.[2] Drive intensity distributes into the affective tonalities of conceptual-feeling. When will becomes desire, impulse assumes a focus as the subjective aim to objects of interest. At a distal phase, the will specifies into a diversity of objects that sap the force of the initial drive. The greater the diversity into which the will distributes, the less the feeling of volition. Free will is diminished by too many choices (Williams, 1981). When the dominant focus in the trajectory from self to object is choice, and there is indecision or a surfeit of options, consciousness of volition may intensify at the cost of a weakness of will.

Generally, choice boils down to two or three options. Possibly, the mind cannot entertain more than a few options because of span limits on attention and short-term memory. Most people have difficulty with two competing ideas at the same time. Choice as a fork in the road requires an emotional tug to incline the self in one direction. When choice is not obvious or outcomes are uncertain, logic rarely carries the day. Self-interest, pleasure and adaptive values account for most decisions, save those when indifference makes the outcome arbitrary. Predisposition, emotional bias and a limit in the number of options are not inimical to free will. Hume noted that action with only one option can be considered freely chosen if inaction is the alternative. Any solitary action has negation as an implicit choice.

Within the psyche independent of brain there are problems with free will and the conditions under which it is presumed to occur. If we assume that the experiential history of the individual is relevant but not dispositive, and that free will obtains in situations when competing choices are limited, the outcome of a decision will be limited by the

---

Agency is the presumption that the self causes action, and free will is the volitional quality of choice.

[2] Similar ideas on the nature of the self, the will and "ascending levels of conative development" were discussed *inter alia* by Stout (1921ed).

options. The individual may be conscious of forces driving the occasion, but among the options, is decision predetermined by mental or neural causation? The problem narrows down to the micro-temporal "anatomy" of self and choice. A fair description of the psychology of choice can then be considered in relation to brain activity, whether brain activity follows causal laws and how to conceptualize conscious choice independent of brain.

Psychology of Choice

Few accounts of choice have theoretical appeal. One work on free will that abandons all insight (Wegner, 2002; review in Brown, 2003) avoids the tangle of consciousness and choice by postulating a causal basis for automatic action through a hypothetical system or pathway and a parallel signal transmitted through another pathway to inform consciousness of the action that is taken. This leads the individual to believe s/he consciously acted even though the action is automatic or forced. Presumably, animals that make choices — a rat in a T-maze, prey selection in higher mammals — do not have such a secondary system or pathway. An hypothesis in which human consciousness depends on a pathway to consciousness for automatic or unconscious behavior, i.e. the *ad hoc* postulation of a consciousness-system that oversees the automatic, is a bizarre speculation without justification, e.g. on the origin of the system, how the systems interact, relevant brain anatomy, evolutionary and maturational aspects, the role of thought, judgment, deliberation, emotion, reason or irrationality in relation to automaticity, the feeling of agency, the value or evolutionary import of consciousness of action without causal efficacy, and the shift from automatic to conscious or from spontaneity to deliberation, and the reverse. The decantation of complex phenomena to neural mechanism, especially the imputation of hypothetical pathways, is the crudest form of interpretation.

The feeling of volition does not require action but occurs with certain forms of imagery. One can voluntarily imagine a mouse crawling on the back of an elephant, decide to think about a topic or call up a memory, a plan, lecture or journey, all in the absence of action. The self is active in relation to some images and passive to many others. There may be uncertainty if nocturnal jerks of the legs are voluntary or involuntary. Stimulation of motor cortex gives limb movements that do not usually feel voluntary. Chorea may have the same quality. In these cases, agency is uncertain, perhaps because choice is not engaged. The only credible theory to date based on neuropsychological data and explicated in some detail is the microgenetic account of choice as an accentuation of an embedded phase in the transition of potential to actual (Fig. 9.1).

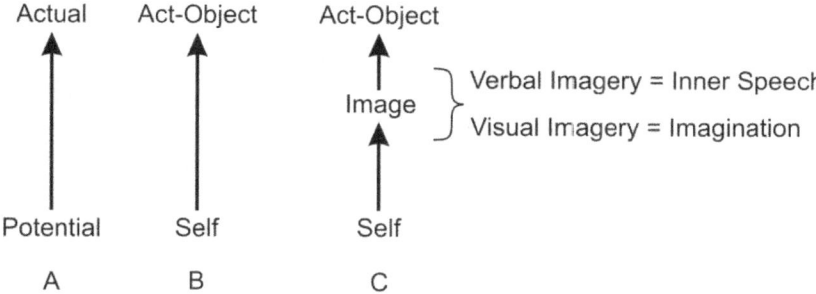

Fig. 9.1: The mind/brain state. A) The transition from potential to actual is depicted from the onset of the state to terminus, but iterated at every phase; B) The same transition leading from self to act and object; C) The image is an arousal of a pre-terminal phase that is traversed automatically in B. Verbal imagery is inner speech or verbal thought. Visual imagery is visual thought or the creative imagination. Dream occurs when sensory constraints are insufficient for the substrates of imagery to realize an act or external object.

Consciousness is the relation of self to its implementations in acts, objects and their precursors in thought. When embedded but virtual phases are accentuated, i.e. actualize, the *implicit* (unconscious) choice buried in spontaneity is revived in *explicit* (conscious) choice. A commitment, aim or specification of one course of action is held in abeyance as the potential for other outcomes is uncovered. The uncovering is an elicitation of novel items from the categorical ground of discarded options. To access the category entails a withdrawal from a perusal of choices to the potential behind them. This requires a postponement at penultimate phases. An occasion of introspection[3] depends on a prolongation (neoteny) or uncovering of the implicit choice concealed in automatic passage.

A common assumption is that introspection and conscious decision are higher-order events superimposed on automatic behavior, i.e. additions to the repertoire of early cognition or the animal inheritance mediated by neocortical regions, not only with regard to free will but for other instances of the implicit/explicit dissociation in memory (Schacter, 1985), in knowledge and action (Isaacson and Spear, 1982),

---

[3] The attempt is to explain introspection, not argue that introspection is explanatory. With respect to neoteny as the basis for introspection, the microgenetic account has affinities with the "perceptual replay" of Lyons (1986).

blindsight (Bender and Krieger, 1951), and other phenomena.[4] This idea resonates with common sense and is consistent with the fact that, in humans, the primate inheritance is refined and supplemented by novel capabilities, especially language.

A central feature of introspection is the *intrinsic* relation of self to mental content, at times passive as in dream and creative thought, at other times active.[5] The self is not in juxtaposition to content as a cause of action or inner experience, but is felt as volitional in relation to the content generated. The self inheres in the objects and thoughts to which it gives rise. Agency is often contrasted with automaticity, but there is a continuum from an extreme of passivity in scenic hallucination or the detachment of object perception to volition and/or effortful deliberation. One moment there is a feeling of volition for thought images, the next the self is passive to a train of memories. When a deliberative act becomes automatic, e.g. when neural formations entrained in the effortful learning of a piece of music are traversed spontaneously in performance, phases that were formerly deliberative, now enriched, revert to automaticity. The performance can again become deliberative if latent possibilities are explored. However, choice does not necessitate options, only the potential for them. The options that count in agency are often indistinct, more like proclivities or tendencies. This applies to action in opposition to inaction or to decisiveness when alternatives are weak, ignored or not imagined.

Introspection is a novel acquisition that is a branch of an earlier stage in evolution and maturation and not, except for the chronology of acquisition, a higher rung on an evolutionary scale. The prolongation of the preliminary, i.e. the arousal of imagery preceding perception, is a manifestation of *neoteny*, a retardation in the timing of juvenile stages. The prolongation of the juvenile in evolution and ontogeny corres-

---

[4] A condition analogous to blindsight occurs in other domains of cognition, e.g. audition, motility (Brown, 1985).

[5] The idea of perception as passive or receptive does not take account of psychopathology and transitional states. The self is passive to after-images, though opening or closing the eyes or diverting the gaze can have an effect. Some agency occurs for the eidetic image. Active and passive feeling occurs with memory images. Hallucination may disappear with limb and directed eye movements. Phantom limbs feel under voluntary control. Hypnagogic images are prolonged with passive observation and disappear with an active attitude, e.g. reaching for the image or looking at it. Inner speech can shift from active to passive, from the *preverbitum* to auditory verbal—even command-hallucination. Within imagery and independent of action, the closer to the perceptual endpoint, the greater is the passivity or detachment, while the closer to verbal thought or inner speech, the more likely a feeling of activity or agency.

ponds to the prolongation of the preliminary in cognition or microgeny. There are no additional structures or processes in introspection, only a prolongation of earlier phases in the elaboration of form. Goethe had the germ of this idea when he wrote that propagation by cuttings shows that the root is everywhere.

What is the nature of the freedom in free will? Is it the realization in action of the will or desire in choice? Is it the feeling of agency regardless of how acts arise? From an externalist standpoint, freedom as a social or political act is often confused with free will in psyche. There is the distinction of freedom to act once choice is decided, and the freedom to decide on that choice, the former ostensibly between mind and world, the latter intra-psychic. If options arise spontaneously, and choice is restricted to the options given to the self, the issue is not the options, which are not aroused by the agent, or the fact that options are spontaneous, but the feeling of choice among them.

There is a sense of "temporal distance" between a self and a decision to choose and to act, and if to act, to decide among possible outcomes. The microstructure of choice reveals many "decisions" — which option, which action, which outcome, when, where and so on. What role do options play in final choices? If options develop out of unconscious predispositions that underlie the self, what part in decision-making owes to the constraints on the self, on options or on their unconscious origins? Can we imagine a self dissociated from its precursors, from prior selves and antecedent experience? What does it mean to freely effectuate an action on options that are spontaneous or the outcome of deliberation? These questions are not resolved by relegating the self and agency to illusions when an account of these phenomena is obligated by any serious philosophy of mind.

From onset to completion in the mind/brain state, there is progression from possibility to specificity or generality to definiteness, a transition shaped proximally by internal constraints of character, habit, belief, value and the preceding state, and sculpted distally by sensation. Transition in the mind/brain state is epochal and modular. Phases exist *in statu nascendi* until the state actualizes, at which point the succession of traversed segments takes on temporal existence. The succession is not a causal motion from point to point; rather, *earlier phases in the state become later ones.*

On Causation and Illusion

The problem of free will is generally framed in the contrast of psyche and brain, which for some is the distinction of illusion and causation. If the self is illusory, so is free will, but an illusory self still requires an explanation of the felt effectuation among mental entities — processual or substantial. This would be non-objectionable to determinists if the question is not whether an illusion can effect change, but how change is effected by the substrate on which the illusion depends. However, the effectuation of acts by a self (or by thought, consciousness) assigns causation to the psyche for self-initiated action. Agency is the presumed causal interface between self and action, evoking speculation on properties of mind that are essential to an action caused by psyche, but are inconsistent with the independence of the psyche from causal (mind or brain) process that is essential to free will.

Put differently, if the self acts in a way that exhibits causal properties, i.e. agent-causation, causation among psychic contents is possible. Free will is vitiated if mental states are caused by prior mental or brain states. The absence of an awareness of causal ancestry, and the feeling that the self is a cause of action, support an intuition of the self as the sole progenitor of a forward-looking causal sequence, not a link in an ongoing causal chain.[6] We do not feel, or believe with conviction, that the self is an effect—a repetition—of the immediately prior self; rather, it persists, with glacial change, as a mental solid. Perhaps this "causal persistence" is explained by the revival of the self in each mental state. The self as a recurrence is not a direct effect of the self of the prior state.

The brain as part of the natural world is a physical entity presumed to obey causal laws. On the mental side, the result is that free choice is problematic, if for no other reason than the implausibility of an effect of an immaterial self or consciousness on a material brain or other mental contents. On the brain side, causation is an article of faith. Yet free will seems real and essential to everyday decisions, as well as in ethics, jurisprudence and social intercourse. When classical science (not quantum theory) is anterior to speculation, all roads lead to determinism. The assumption of causation in brain and the asymmetric relation of mind and brain[7] tip the scales against freedom of choice, with options, judgments and actions hostage to causal brain activity. Pears (1963) puts the matter clearly: "the route of the physical determinist's explanations would surely have to run via neurophysiology." Indeed, the assumption of reducibility to neuroscience is so widespread that one of the few options to avoid causal inevitability and the elimination of

---

[6] The escape from determinism in free will is analogous to the escape from *Samsara* in the cycle of death and rebirth.

[7] Few would argue, except dualists or those who believe in a soul, that mind is independent of brain, but might brain be independent of mind? The way to pose this question is with the pattern of the brain state that corresponds to that of a mental state. It is doubtful such a brain state could exist without a mental state.

mind, quantum indeterminacy, is invoked as an escape to probabilities that, so it is argued, open the door to non-causality. However, choice as probabilistic would result in random or unpredictable decision, whereas the selection of options—the active feeling of the self in relation to mental content—is a feeling of relative certainty in action and decision-making.[8]

Thus, free will is burdened on the brain side with a presumption of causation, and on the mind side by the presumption that free will is unreal.[9] If mind and brain are dual aspects of a common process, if brain activity follows causal laws, and if mental states depend on brain states, free will is not possible, for there will always be a causal brain process underlying and preceding every mental phenomenon. On the other hand, if free will is real and still corresponds to brain activity, the brain state could not follow causal laws. The problem for free will is not the reconciliation of psyche with causation but of causation with brain process.

When the focus is on neuronal connectivity, the case for cause and effect seems relatively unambiguous. But a brain state enfolds a fluid configuration in relation to the synaptic strengths of myriad neurons. Passage over phases has the character of a traveling wave with constraints at multiple segments. Causation in brain and determinism in nature are not facts but hypotheses. An inference of causal determinism is not based on brain study but on extending to brain the classical laws of physics and the assumption that choice and action correspond to brain process, that the self is deluded into thinking the decisional process is free and that an alternative choice might have been made. But the assumption of causation is at odds with microgenetic theory, in that the passage in a mind/brain state from inception to actuality means that an early segment does not cause a subsequent segment but transforms into, i.e. *becomes* the subsequent segment.

Many people accept brain function as causal and free will as illusory without an account of the mind/brain state. There is an avoidance of such questions as, if free will is an illusion why do we have it, why it is so powerful, how it comes about, and what justification is there for the idea that brain function is grounded in physics and the psyche is purely

---

8   It is conceivable that decisions are probabilistic but interpreted by the agent as willed, especially since the onset of a voluntary action precedes consciousness of the decision to act.
9   Free will is no less illusory than any other mental phenomenon, including the belief in a self, in substance, time or the reality of the perceptual world. The entire mental life is illusory, yet free will more than any other illusion is singled out for special consideration because, unlike other illusions, it violates the materialistic belief in universal causation and determinism.

phenomenal. Free will is described as illusory more readily than other phenomena because choice conflicts with causation, while other illusions — objects, substance, action — are accepted as real because there is no explicit conflict and doubt is largely theoretical. Free will is a special type of illusion that merits this status because it violates the dogma that brain function is causal.

What does it Mean for a Phenomenon to be Illusory?

There are many different senses of illusion and many different phenomena that are termed illusory.[10] The entire mental life can be considered an illusion (Vaihinger, 1924/1965) or in Indian philosophy, *Māyā*. Some phenomena we recognize as illusory, e.g. duck/rabbit figures, rainbows; for some we are largely unaware, e.g. perceptual constancies, time experience; others we accept as real experience, e.g. dreams; while other phenomena, especially acts and objects, are difficult to accept as unreal. Yet, an object is an image extracted from binocular disparity that depends on memory (temporal lag), while action is experienced through recurrent collaterals as a perception and also shows a temporal lag (readiness potential). Conscious acts and objects are virtual images, off-line with real world correlates.

The realness of some phenomena contributes to this judgment — objects are more real than thoughts, thoughts more real than dreams (Brown, 2004). Phenomena such as object perceptions are so real we have difficulty believing they are neural configurations, not solid things in the world. We confuse the feeling of realness for reality when, in fact, we have no direct knowledge of reality in spite of the realness of experience. For some, consciousness is real, but anchored at one end by a self that is presumed to be illusory, and at the other by an illusory world. Hallucination may seem real with no correspondence to external events. Indeed, the reality of certain hallucinations, or dreams, even when the image is out of context with perception or memory, is a sign of the ingrained sense of realness for mental imagery. The feeling of reality for a common object — the veridicality of object representations — is a strategy for survival, namely, to reliably map perception to the world. A person who thinks, and more importantly believes and feels, that a tiger is a mental image would not be around long enough to reproduce.

---

[10] In neurology, illusion is generally contrasted with hallucination, not with delusion, which is closer to its use in philosophy. In some sense all of these phenomena are false beliefs, though only delusion is uniformly accepted by the individual as a true belief or real event.

Some phenomena—acts, objects, truth-statements—map to an inferred reality, others such as dreams do not, but most illusions have one foot in psyche, the other in the world. The self is not only taken as real by most people, it is our most precious possession, that which enjoys pleasure, gives pleasure, defines us, suffers, dies. Having a self allows us to infer or deny it in others, and in animals. Most people believe that personality distinguishes one dog from another and is a sign of subjectivity, even the *anlagen* of a rudimentary non-reflective self. We give names to pets, especially horses and dogs, treat them as friends, talk to them, mourn their loss, and impute to them a near-human nature.

Duration and the "Illusion" of Self

Acts and objects develop from unconscious to conscious over some duration, perhaps a half-second or so. The temporal order is that of passage from earlier to later. This is not the virtual duration of the present that is nested in the disparity across segments. Early phases of personal memory, implicit beliefs, values and instinctual drive categories lead from the inception of the mind/brain state to the self at the threshold of consciousness, to conceptual-feeling and conscious behavior. An awakening of the image-field fractures potential to final definiteness. This passage from self to object, from desire to idea, realizes a subjective aim, which is purposeful or teleological, unlike causation and determinism.

The duration of the present is extracted from the <u>temporal</u> disparity between a state of perception and one of prior revival as a transposition of the relation across segments in mind to an external series from past to present, i.e. taking a one-dimensional point to a two-dimensional line (Bergson, 1923). In contrast, objects are images that are extracted from <u>spatial</u> disparity (binocular perception, which takes a two-dimensional flatland to a third dimension. One disparity expands temporal order with duration; the other expands space with depth. Duration is to time as depth is to space. Agency arises in the transition from self to act experienced as an internal sense of motion and serial order. The individuation of a self in opposition to image and act depends on temporal disparity. What is real or illusory in mind must correspond to what is non-illusory in brain, the most likely candidate being an iterated whole–part relation that specifies options-as-actualities and individuates the self-as-potential to diverse acts and objects. The outcome is an inner theater felt as real in spite of its illusoriness, as if multiple illusions in an illusory present conspire to produce a reality that is as real as it gets.

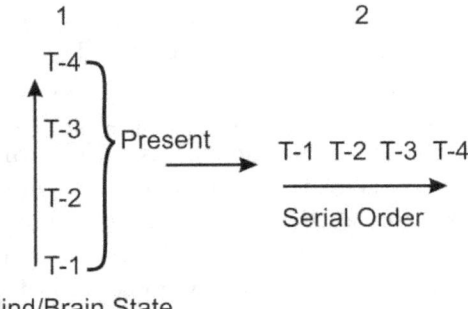

Fig. 9.2: Mind/Brain State. 1) The "stacked" sequence of events in the mind/brain state is epochal and simultaneous. The passage from before to after (T-1 to T-4) is an objective time-series. The specious present is extracted from the disparity between T-1 (a prior state) and T-4 (the current state). 2) The before/after is transposed to a sequence from past to present in the duration of the now. Serial order is apprehended as a "line in time" from past to present, *in the mind* such as a melody, or *in the world*, such as a sequence of events in visual perception (Brown, 2010a).

The self–act relation and the duration in which it occurs, i.e. the specious or phenomenal present (James, 1890), is a virtual arch over passage in the world and, by inference, the brain, i.e. the duration spans a succession of phases in the brain state as well as passage in the world. If free will involves the "detachment" of mind from causal process in the brain, duration may clarify the difficulty. The succession of before/after in the mind/brain state in a simultaneous epoch (Fig. 9.2.1) is transposed to a series from past to present (Fig. 9.2.2). The present shifts physical to subjective time. The bubble of the present transposes unconscious passage to conscious event-order, permitting a mental life not possible on the knife-edge of change. In the before/after an instant dies as the next appears. In an epoch, before and after are simultaneous until the state actualizes. The duration that arches over succession in the mind/brain state becomes the duration that enfolds succession in the passage of the world.

Remarks on the Psychology of Free Will

The feeling of what is real does not correspond to reality (e.g. the realness of hallucination); nor does the feeling of freedom necessarily correspond to a freely-decided act. The consensus that much, if not all, conscious mentality has an unconscious origin is alone sufficient to dispel the notion that the self is independent of its ongoing and experiential history. The temporal lag in perception and action refute direct on-line perception or perceptual immediacy. Self and thought are grounded in patterns that have adapted to occasions of individual experience. The problem is three-fold.

First, any decision or freely-chosen course of action can be assigned to brain or, retroactively, to some conscious or unconscious bias. Even an act that defies self-interest, such as altruism, or one that differs widely from the mean of prior decision, or seems wholly impersonal, can be construed as determined by some cause(s), known or unknown, or concealed tendencies. Indeed, psychoanalysis makes a profession of filling in the blanks of unconscious causation. Yet, this is causation by implication: that brain is causal; that antecedent history is causal; and that consciousness, the self and its contents are illusory fluff, while the real work is done by the brain.

Second, the inability to predict many acts, which is true for spontaneous and deliberative action, induces an uncertainty that appears to justify free will. But uncertainty can be attributed to the probability of outcomes, not the enactment of deliberate judgment. A lack of knowledge of the future or the aim to bring about certain eventualities can as readily be attributed to openness of choice as to causal closure. Choice as a surprise to the agent, e.g. a decision that "pops into the head", or the "bubbling-up" of decisions aroused by a complex and largely unconscious set of conditions, is interpreted as evidence of a causal chain in which the self is ingredient but does not freely intervene.

Historically, the impersonality of reason has been the measure of free choice, but most decisions, though reasonable, are not motivated by reason. An act driven by emotion, or one in which emotion is the essential factor, is not necessarily less free than one decided by reason, since one can weigh conflicting emotions as one can judge competing reasons. Emotions are less well-defined than rational thoughts and ought not, though they do, participate in rational decisions, particularly in experimental settings such as deciding when to lift a finger, though there is awareness of an impulse that sets the action going. Everyday activities, such as getting out of bed in the morning or turning off the radio, often feel spontaneous; they are not quite emotions or ideas but rather impulses that activate movement. One could say, impulse is a momentary dominance of the affective quality of a latent idea.

Some might argue that action stemming from emotion is external to reason or intellect and taps into genuine character. Is emotion less impersonal or is the bias in reason better concealed? Lying, deviousness, selective recall or rendering of fact are characteristic of reason not emotion. The effort to suppress emotion even when it is the primary motive can lead to appalling judgments, especially when empathy is trumped by grandiose strategies or beliefs. To profess concern for too many people is to exchange a genuine emotion for an arid concept.

Reason can justify almost any action, or demonstrate that an evidently false belief is rational, while emotion, so we believe, generally runs true to character. But in the end, reason and emotion, concept and feeling, being and becoming, are part of the same construct, different racks on which to hang one's opinions.

Third, as discussed, the onus tends to be on demonstrating free will, not on documenting causation, since cause and effect are assumed to apply universally in scientific, social and psychological discourse. The irony is that while mental phenomena are assumed to be illusions, or in any event phenomena with a causal basis, causation in brain function is an hypothesis that stems from a conceptual apparatus that is, itself, deemed illusory. That is, the assertion that behavior is caused by the brain, or that mental phenomena are reducible to brain and constitute a causal series, is a reasonable—though not necessarily correct—judgment, but the assertion depends on a rational consciousness, which for many is an illusory effect of brain function. In a word, the paradox is the affirmation of the "reality" of causation by the "illusion" of conscious thought.

While science has the value of canceling error through verification, true or false statements are still products of self, choice and decision. If thought is caused by the brain, there should be causal insularity. Thought should be non-generative and predictable. If mentality is an illusion, how can we ever know that thought is true or that it describes real events? How can a judgment be true if the self that makes that judgment is illusory? If the causal theory of brain or the judgments that lead to this theory are themselves products of brain process, how does the corrective function of verification intercede in a presumably determined sequence? What is the status for mind/brain theory of imagination, dream, creative thought, psychosis or radical ideas later shown to be true?

On Constraints and Partitions: Toward a Theory of Free Will

For microgenesis, the *pattern* of activity in the brain state is not a boxcar succession but the elicitation at successive points of fractal-like parts from category-like wholes.[11] The parts are qualitative specifications, not self-similar replications, while the wholes are categories or

---

[11] Theory of chaos and fractals does not relate directly to the problem of free will. The point here is to distinguish the self-affinity of fractals from the qualitative shifts in whole–part transition (MacLean, 1991, and other papers in that volume).

potentials, not sums or containers.[12] Every partition has a whole that individuates the parts. This raises the question, if the self is a whole or category from which ensuing derivations arise, does it, by virtue of its locus, exert an influence on the cascade of particulars? Does an endogenous whole–part process guided by constraints qualify as causal? Constraints on developing form are analogous to the effects of a river bed or bank on the pattern of flow. Is this causal? If the progression over phases in the mind/brain is an iterated partition of an initial configuration, such that the configuration at onset and final partition constitutes the derivation of a single form, not a sequence of causal pairs, in what sense is this progression a causal series? To stay with the analogy of a river, or a wave that rolls to the shore, we are dealing with early and late phases of a single form in change, i.e. a becoming, not a series of causal pairs.

A limitation of possibility differs from a causal effect, primarily in the absence of contact of the constraint on the surviving form. Constraints limit possibilities but do not cause residuals. For example, if there are four possible paths but only one is taken, and if the likelihood of three paths declines with inhibition, lack of facilitation, etc. while the fourth is untouched, the inclination to take that path appears forced, even inevitable, but is it caused?

If wholes that partition can be said to control the parts, this might imply that the self, as a foundational whole or categorical prime, constrains and implements its individuations. If beliefs and values in the core of the self exert an influence on final actualities, the ideas, objects and multiplicities that individuate can be viewed as the subjective aim of a process of self-realization. The configuration that develops out of the self is modulated at successive points by sense-data to elicit an adaptive resolution. The significance for free will is that acts and images specify implicit attitudes through a system in which personal belief is exposed in explicit choice. Contents in consciousness define the self that instigates them. Choice apprises the person of the parameters of decision, but decision traces to an individual self so that choices and acts will be the outcome of personhood shaped by necessity.

Are the conditions for free will satisfied if one is conscious of options that arise from dispositions even if the process leading from disposition to option is unconscious and the sequence corresponds to

---

[12] Mereological studies of wholes and parts tend to be synchronic and substantialist, with parts as constitutive and wholes as aggregates (see Bacon *et al.*, 1993). In a mereology of transition, the virtual parts of a non-articulated whole are elicited in a qualitative becoming. Identities are not individualities but category-members, which are themselves categories to further partitions.

brain process? Free will entails the ability to decide among a set of options according to the interests of the agent, not openness to every conceivable possibility or disinterest in a course of action extrinsic to personal dispositions. Decision adapts need and desire to the social or natural environment. If free will is illusory because it depends on brain, what of all other internal or external phenomena that also depend on brain? If free will is based on choice, is there implicit or explicit choice in every act of perception, e.g. when, where, what, why to look at an object, how long, what features? Choice is consciousness of exposed irresolution at penultimate phases in psyche or brain.

The notion that choice applies to conscious choice, not contrast or possibility, i.e. not the genesis of self and thought, takes the problem outside the mind where it can be explored and debated like a truth-judgment in which antecedent process is inessential to the matter being decided. Free will is not necessarily vested in *conscious* choice, but instead is in the process leading to options before or after one is conscious of them. Choices appear in consciousness but acts and decisions based on those choices develop unconsciously. The possibilities that count in whole-part transformation are antecedent to the choices. Once possibility is conscious, as an idea or image, its role in the process that underlies choice is relinquished. To the extent we have free will it is not in choosing between conscious options but in the micro-temporal development of self and options as they become conscious.

Take two conscious options, whether to dine at one restaurant or another. The individual may consciously review the price, menu, ambience and so on but, *ceteris paribus*, when a decision is made, it often feels more like an impulse than a rational analysis. Is a decision between conscious options more or less free than a decision owing to passage from self to act after options are contemplated? We may think the conscious self decides, but self, choice and the process leading from self to choice are inaccessible. Whether or not options are conscious, decision and action constitute distinct mental acts — deciding and acting — not scrutiny and contemplation. Thought serves to unpack ideas latent in unconscious categories. Every mental content, every set of choices, develops and is sustained over many mind/brain states. Options are a foretaste of the implicit choice through which decision occurs. The problem is, if decisions are not fully conscious, in what sense are they free?

Constraints have a long history in philosophical and religious thought, in the *via negativa* of argumentation as well as in science, as in the style of Sherlock Holmes that when all possibilities save one are eliminated, that which remains must be true.[13] The closure of possibilities is the other side of the partition of wholes. Constraints affect options unborn, not the perusal of clear choices which seem to many the criterion of volition. Conscious choice is always a venture into possibility.

Of relevance are studies of the readiness potential by Libet (1999), which have been interpreted as demonstrating that consciousness has a "veto" at the final stage of a developing act. The work shows that a voluntary act begins well before the agent is conscious of choice. However, the veto is not a negation of all surfacing possibilities but of one or two finalities, and thus tantamount to a tacit permission for a preferred —usually, the most adaptive—outcome. In my view, *a conscious veto is a final stage of constraints that influence all segments in the forming act and object. There is a continuous reconciliation of possibility and specificity, though we are conscious of only the distal portion. The final act or object is the terminal specification of background context, the potential of which translates an implicit choice to a final actuality. Options are not waiting in situ for activation, but become known as they become conscious. The problem of free will rests on the individuation of choice out of potential, or item out of category, as parts resolve from wholes by veto or constraint, and the degree to which the self controls the specification process.*

Since parts are not constituents but qualitative individuations, and since no entity is exactly the same on different occasions, the whole/part shift is better described as novelty in defiance of causation. We speak of creativity when novelty arises from a conceptual depth. Bergson referred to creativity in relation to freedom, writing that the whole thrust of evolution is to insert indeterminism into matter. Bergson (1920ed) constantly emphasized the relation of duration to freedom, for example, that *"the duration of the universe must be one with the latitude of creation which can find place in it"* (his italics), a relation elegantly captured in his *Time and Free Will* (1910). Whitrow (1976), discussing the time series of McTaggart, also invokes duration, in which "time is the mediator between the possible and the actual."

The concept of actualization central to microgenetic theory entails creativity in mind/brain process and the passage of world process as well. Given that parts achieve definition as they actualize, and serve as wholes for further transformation, novelty is inevitable. In writing of determinism in concrescence—which bears a similarity to actualization in microgenesis—Whitehead (1978ed) noted some determinism in the subordinate process but added that there is always a remainder for the

---

[13] Holmes put it this way: when I have eliminated the impossible, whatever remains, no matter how mad it might seem, must be the truth.

decision of the subject. In referring to whole and part, he went on to say that the "final decision is the reaction of the unity of the whole to its own internal determination". For Whitehead, the transition from generality to definiteness in an act of concrescence arises out of Creativity as the foundational principle of the natural world.

Theoretical Note One
Probability and Potential

> *O chestnut tree, great-rooted blossomer,*
> *Are you the leaf, the blossoms, or the bole?*
> *O body swayed to music, O brightening glance,*
> *How can we know the dancer from the dance?*
> —W.B. Yeats, *Among School Children*

The passage from potential to actual, first described by Aristotle as a tendency or entelechy in the realization of possibility, has figured to a varying extent in many philosophies, including that of Dewey, Bradley, Whitehead in particular, and others, as well as microgenesis. Cobb (2008) describes Whitehead's views as follows: "a proposition is defined very much as an eternal object is. The difference is that an eternal object is a *pure* potential and a proposition is an *impure* potential. An eternal object is disconnected from actuality; a proposition is tied to it. Propositions come into being along with actualities. Eternal objects do not. They are strictly timeless." In spite of its occasional impenetrability, Whitehead's thought on this topic is similar to that discussed here, namely, a progression from potential as abstract and timeless—I would say, having the non-temporality of the unconscious—to a concrete (conscious) actuality.

Popper discussed potential as a kind of potency, proclivity or predisposition, but the idea of potential is close to that of wholes and parts, or the generation of diversity out of unity, or multiplicity out of oneness, or the passage from generality to definiteness. The critical aspect is the elicitation of contents or member-items out of background categories in a transition from the abstract to the concrete, or from the virtual to the real. The idea resonates with the shift from intuition to action or from competence to performance. It can be applied to the relation of belief and doctrine, or receptiveness and dogma, yet the topic is shrouded in mystery. Partly this is due to the ill-defined nature and uncertain properties of potential. Does potential *contain* the multitudes to which it gives rise, and if so, are its properties a function of its constituents? If not, how can its properties be characterized? Related to this is the assumption of wholes as sums or containers and parts as embedded elements rather than latent possibilities.

The difficulty, as Dewey recognized, is a shift from diachronic to synchronic thought that plays out in the exchange of potential/actual as time-creating to cause/effect as time-dependent. As a category, potential is non-temporal and becomes temporal in the actualization of definite particulars. This entails an incorporation of distinct elements with causal properties, for wholes resist characterization unless the parts are specified. The inability to specify the parts of antecedent wholes is to be expected in that parts are virtual or possible derivations that become real when they can be designated and competing parts are no longer generated. The "linkage" of timeless wholes to temporal parts is through a procession of categories that are striving for finality, i.e. toward a subjective aim.

Unless a derivation actualizes as an act or object, it forms the ground for further partitions. Only when the parts achieve closure in final objects, as when a concept becomes an object, do they take on real physical (causal) properties. Specifically, the individuation of categories in the mind/brain state is a succession of virtual items developing as contrasts in the category. This is not a series of contents that proceeds step-by-step to a conclusion. Intervening phases are transformed to later ones unless they actualize as contents (images). The invisibility of the transition gives the false conclusion that outcomes are defining features of antecedents in a conscious recombination of traces nested in the potential from which they originate. Specifically, in modular or copy theory, acts, objects and/or their constituents are assumed to be tacitly represented in source-potential. There is no process of realization from whole to part, only an uncovering of parts imminent in wholes.

Gradually, the concept of potential has been discarded for a theory of object- or action-formation in which an unconscious cause is linked to, or identified with, a conscious effect. The outcomes of preparatory phases—objects, statements—are analyzed into elements that forecast or induce them, or the rules that govern the translation process. For example, the features of an object—color, size, movement—are presumed to be constituents in its assembly. The procedure is to relocate object-features from the external world to the mind/brain as formative elements. Here, there is a like–like reduction of the external to the internal or to an earlier phase in mental process. More commonly, the reduction is to lower level physical entities such as individual neurons, columns, distributed systems or genes. In the case of the genome, there is the potential for a variety of phenotypes depending on innate tendencies, environmental conditions and other contingencies, but in principle, potential can be specified other than as a predilection, e.g. in genes, gene combinations, inter-genes, timing.

One field in which potential to actual is fundamental is psychoanalysis, which proposes that early-acquired, unconscious and causal precursors of conscious thought and action can be recovered by interpreting behavioral effects. On this theory, the potential for conscious ideation or action consists of unconscious memories of events in childhood preserved much as they were experienced. The potential of these events, or their neural correlates in experiential memory, is less the possibility of diverse later outcomes as a directive from unconscious templates. The capacity of potential to effect behavior in maturity assumes a degree of certainty in light of the similarity of pattern across individuals. The expectation is that understanding the causal relation of early events to mature cognition will loosen the grasp of an unconscious script, introduce novel possibilities to rigid causation, mitigate necessity to disposition or proclivity and allow reason to replace repetition or habit. In this way, variation (freedom) can emerge in the potential for adaptive action. This depiction of psychoanalysis may not be apparent to its practitioners but it appears to be a principle justification for the method.

In many respects, the relation of whole to part or potential to actual is like that of category to member. The analogy of potential with category, and parts with category-members, conforms to the idea of a category as a grouping of virtual items based on shared features. The implicit category-members are not contained *in situ* but are derived from the category through a process of specification. For example, while the category of animals is essentially limitless, it can be parsed to sub-categories (dogs), then to sub-types (terriers), then to individuals (Fido) with the potential to actualize or undergo further partition, whether a physical feature of Fido (leg) or an attribute (loyalty). All mental contents or events—motor, ideational, verbal, affective—are specifications of categories. The virtual is progressively delimited until it deposits in content—abstract or concrete—at which point the source-category is abandoned for the sub-ordinate. This unconscious process goes on in every mental state. When an item in a category is selected, as when animal is derived to dog, or dog to terrier or a specific dog, the wider category, though ingredient in the particular, is replaced and, in its revival, forms a generative background for further specification.

Object- and lexical-categories have fuzzy boundaries and overlap with other categories, as do members. The category is not an inventory of members, which do not actually exist unless they individuate. The progression is from inclusive and abstract categories to exclusive and narrow instances. Dogs exist as possibilities, like unicorns, but unlike unicorns, dogs *can* exist as actualities, but they exist from a certain perspective as concrete particulars. Even Fido can be parsed as a category of parts in space and time, snapshots of motion, of aging, a moment of interest, an instance of value. Fido is a category of appearances. The sameness of a particular depends on perceiving it as a category of instances, while its reality depends on the observer accepting it as a finality.

The category-member relation occurs in all modes of cognition, but word-categories are perhaps the clearest examples. Behind every word, act or object is a category from which the item arises, to which it belongs, and in which it stands in relation to the properties of other parts or members. Observations of the pathology of language show a zeroing-in on target items through categories of increasing specificity. Similarly, motility can be described in terms of oscillations in which there is a fractionation of a fundamental frequency to a derivative or harmonic. In this process, the proximal is prior to the distal as support and precondition. Limb movements are elicited from postural sets or axial movements such as walking. The older, more primitive enfolds the recent in evolution and in cognitive process. An object-concept is a class of virtual objects related by shape, use or other features. Focal attention develops out of diffuse perception. Foveal vision develops out of ambient vision. Bradley's hesitant flight from an "unearthly ballet of bloodless categories", and the embrace of categories in Buddhist thought, posit a series from abstract to concrete in a hierarchy that is essentially bottomless.

Probability and Causation

The concept of constraints on developing form, the traversal to actuality of abstract wholes through incipient particulars, the arguments against atomic elements in potential and the idea of categories as the ground of novel specifications raise the possibility that *parts are what they become according to probabilities in the traversal of a configural wave*. In a copy theory, particulars are accessed or retrieved into consciousness without change other than recombination. Among the reasons to doubt such a theory, which assigns stability to the trace and dynamic to consciousness, is that it presumes all possible responses to the surround are encoded in a pre-established set. In a qualitative transformation, partitions allow the organism to adapt to novel situations. If the transition from potential to actual is contingent, not causal, elements would not be subsumed in potential but would "evolve" over phases to achieve a balance of need and exigency. The derivation is a continuous shaping at each phase of what will become the subjective aim. While the process is probabilistic, the outcome in speech or behavior is *felt* as spontaneous or voluntary, but if the latter, not as a probability but a causal effect of an agent. Causal feeling in action is from mind to world; in perception, from world to mind.

The hypothesis advanced here is that adaptation occurs when possibilities are whittled away in the realization process and what remains is the most probable outcome. In spite of the probabilistic nature of the process, there is *felt* certainty for agency in behavior and *felt* certainty for the reality of objects. Passage in the mind/brain state is opaque to the actor, who feels his acts, for the most part, are intended or voluntary, i.e. self-caused, and that his perceptions are uniformly independent or object-caused. Acts and the choices that accompany them are not produced out of thin air but are (felt to be) the effects of an agent's dispositions. In perception, sensation is inferred as a cause. The direction goes outward in one modality, inward in the other. Agency and the reality of objects require "blindness" to antecedent process, whether novel, causal or probabilistic. The sense of freedom dissolves if agent and options are causal products or contingent outcomes, and reality is imperiled if objects are felt as mental contents. Agency entails freedom in choice and its implementation, but it equally requires a causal relation of self or thought to action. Ordinarily, we have a choice of our acts but there is nothing we can do about our objects.

The illusion of non-belongingness in perception (mind-independence) that is essential to the reality of the perceived world contrasts with that of belongingness in action (mind-dependence) that is essential to agency. Objects fully detach to form an external space; actions partly detach in a space of the body. This is critical to the feeling of control for actions but not for perceptions. Objects are accepted as real and factual since there is no other reality to compare them with. Hallucinations are recognized as mind-dependent until they replace perceptions, at which point, once again, there is no alternate world for comparison. Action may be instrumental in establishing a distinct bodily space that is in relation to its perceptual counterpart which is taken for physical reality. Agency and freedom are assigned to action and the antecedents of perception in imagery.

A sequence of virtual sub-categories eventuates in actualities contingent on the individuation process. Every phase in becoming is an opportunity for novelty though habit and the elimination of alternatives serve to enhance survival and keep aberrant acts and objects to a minimum. In pathology and creativity, the "road not taken" begins with foundational beliefs and is most prominent at pre-perceptual phases of imagery and implicit choice. At times, this leads to greater novelty, at other times to pathology. The inventiveness of dream owes to a lack of constraint at the perceptual endpoint. Dream is an elaboration of the interior of perception without final sculpting, its images less adaptive and less predictable than in wakefulness. *The alternative to a copy theory, in which parts are subsumed in wholes, is a process in which probability or contingency plays the major role. This allows latitude of options, adaptive flexibility and degrees of freedom in micro-temporal passage. The elicitation of particulars in the actualization of categories or the satisfaction of wholes through probabilities in a momentary state entails that entities are deposited as a collection of transitions, or that change creates stabilities, or that process, though non-causal, deposits causal objects; in a word, the contingency of becoming gives way to the determinism of being.*

From Probability to Causation

The series from potential to actual is non-conscious. There is no content to fill intervening phases other than that which is inferred from experimentation, from normal experience (dream, illusion) and from pathological cognition. The lack of awareness for the process of object-formation reinforces the impression that objects are "out there" and mind-independent. Since the transformative process is unconscious and only final content is available to the individual, who is oblivious to the formative process, we are beneficiaries of our thoughts, objects and acts, which are less our activities than our discoveries. Without direct experience of micro-temporal process, the choice is a passive or active attitude to the final content, to be recipient in perception and an agent in intentional acts. An active or passive stance is less a choice than a concomitant. The feeling of activity (James, Wundt) gives agency; a passive bias is due to the "separation" of objects and the assumption they are independent, or, as in hallucination or dream, when the image is the final phase.

The active feeling in action, and the immediacy of perception, are partly a result of the forward (past-to-present) direction of realization. *Acts and objects develop outward over parallel phases from mind to world.* Agency mitigates the feeling one is at the mercy of the environment. Passivity, which is linked to mind-independence, establishes the reality of external objects. It is essential to feel one's acts are self-initiated, but in perception this feeling leads to the belief one is thinking up his own objects. In schizophrenia, agency for objects, e.g. the confusion of thought and perception, and its loss in action, e.g. that one is a puppet controlled by invisible strings, result *inter alia* from an absence of the deception that objects are mind-independent. Schizophrenia exposes the horror of real insight to the true nature of mind and perception.

In sum, potential and category are equivalent, with parts or members developing in a progressive specification according to the probability of actualization or traversal, with the final content representing constraints on novel form, not anterior cause. What actualizes is what resists elimination. Lack of awareness for transition, forward development and feeling of activity are essential to agency. That agency occurs with pre-perceptual images, e.g. imagination images, inner speech, illustrates that the development of acts and objects is parallel, not reciprocal.

Theoretical Note Two
On Truth

Introduction

The usual approach to truth is to frame a proposition or a statement of belief in a way that maps or corresponds to, or can be tested against, empirical data or has application to conditions in the world. These conditions are preferably impersonal, since a statement such as "it is raining" or "I believe it will rain" can be verified by the observer and others, while a statement such as "he is sad" or "she loves her husband" is less open to validation; it can be questioned by the individual about whom the statement is made as well as the testimony of others. Conflict is equally apparent in the observer, whose own mental states can be as opaque as those of others. A person can give what he believes to be a truth about himself or others that is misleading, self-serving or unjustified. An initial observation is that if we exclude personal statements about one's self or others as being largely indeterminate as to truth, as well as statements about feelings, or those statements not in a form accessible to truth-judgment, for example poetry or metaphor, we have already limited the scope of an inquiry as to true propositions to a relative handful of instances, primarily scientific concepts.

We are accustomed to taking an everyday truth with "a grain of salt", in that the psychology of language and thought, which philosophy tends to ignore, plays a role in choosing, asserting, framing and defending a truth. The motivations and context around the statement—evident and concealed—affect its rationale, as do needs and values. These psychic "contaminants" are split off or discarded as irrelevant to a judgment of the truth of a statement, much as the personality of an artist is cleaved from the truth or beauty of an artwork. The statement is presumed to be independent of its antecedent or concurrent context and treated as a consensual object for debate and discussion. A truth judgment implies a rational statement. In art, the irrational is often viewed as inspirational, for the artist and the observer. In philosophy, irrationality is conceived as an aberration, certainly not a phase through which the rational is achieved.

Typically, a proposition or "interpreted statement" is conceived as a product of a belief, with the problem of establishing the truth of the statement identical to that of the truth of the belief. The relation of belief to statement tends to be collapsed in a fusion of the two, with the underlying belief system presumably constituted of the (true) statements it represents or generates. The problem also has a metaphysical aspect, in that the correspondence of a statement to reality, on which the judgment of truth is based, depends on some form of realism for the (perceived) world. A state-of-affairs in the world is an objective determinant of a subjective belief, whether in the identity of sentence and belief, or as to the acquisition of the belief and its role in sentence production. A difficulty for correspondence theory is that a mental statement or a pattern of brain activity corresponding to a belief or proposition is not identical to actual events in the world. The theory assumes a correspondence between a subjective content or physiological brain state and an objective fact, but if inner and outer, subjective and objective, are different things, what exactly is the correspondence for?

In idealism, the coherence of belief is fundamental in the embedding of world and proposition in a knowledge base, since the world is conceived as phenomenal or as a projection of ideas. The problem of a difference between subjective and objective does not come up, since the objective is an objectification of the subjective, i.e. the world is the same "stuff" as the mind. Coherence of beliefs, or of facts within a belief system, is the primary determinant of truth, but this view has the difficulty that facts are established by experience and coherence must appeal to some states of affairs other than internal consistency. People are entitled to their beliefs and opinions but not to their facts even if facts are the residue of beliefs. A system of belief could be coherent but false, as in animism, delusion or psychosis, without the decisiveness of fact to establish a standard against which a given belief is judged. An entire "reality", an alternate universe, can be constructed in a work of fiction. Since an isolated statement or fact that is discordant with belief will not be tolerated, e.g. that pigs fly, coherence is essential to belief but insufficient for truth. To resolve an impersonal truth, idealism must entail coherence, but also requires conformance of belief to actuality.

Generally the closer a truth-judgment to logic and mathematics, or to local conditions of life and mind, the more readily truth or error is ascertained, while statements of wide scope or depth, or those of personal import such as moral assertions, are either accepted as intuitively correct or rejected as inconsistent with personal experience, fairness or common sense. In most instances, a person judges a statement as true or false according to whether the statement fits with the belief system, i.e. coheres with experience and habitual thinking, and whether the belief fits with external conditions. This applies not only to everyday judgments, or to behavior that depends on implicit truths, such as taking an umbrella on a rainy day, but to topics of philosophical import. To say, "an object is x", where x could be real or unreal, fact or possibility, causal, redundant or novel, object or event, and so on, can incite debate no matter how the sentence concludes.

It can be said that truth-claims are reliably ascertained only in propositions that refer to facts, and that facts relate to beliefs that for the most part, i.e. excepting early implicit beliefs or the animal inheritance, have been previously established, and that observations and facts can only be determined if set in a verifiable statement that can be tested, or if the statement is crafted in such a way as to eliminate disqualifying observations. The statement, "grass is green" is an illustration of such a proposition, but it applies, on the one hand, to a certain type of grass at a certain stage in maturation, in relation to water, sunshine and nourishment, to season and certain conditions of light and shade in a person with normal color vision. On the other hand, the statement is true if it conforms to beliefs about the nature and color of grass on, say, a well-tended lawn, not the African savannah. The objective truth of the statement rests on implicit beliefs about its terms, which are learned and assimilated, and regenerated out of the knowledge base, i.e. semantic and episodic memory, implicit and explicit, that are inculcated by past experience.

In correspondence theory, since the comparison is to external fact, the truth of a statement depends on, or is aided by, consensus, which is a means of overcoming doubt. Coherence theories are not decided by consensus but by consistency, since doubt, in principle, is excluded by coherence. The certainty of a truth does not depend on whether a fact or event is past or present. For a past event, the fixation of what actually happened should tend, logically, to eliminate disputes over the factual basis of the occurrence. This lends greater authority to external fact. Even a presumptive truth about the past is a fragment in a whole truth that can never be known or understood or an approximation to the historical record, and even if the assumption that the past is forever fixed justifies the mapping of a statement to this record, in memory or scholarship, the record is invariably incomplete, often corrupted, and can evoke different interpretations. Even past facts that are indisputable are true only in a general sense, since specificity opens the door to differing perspectives and unsettles the certainty of what is known. This is still more the case for the ongoing present, in which a clash of perspectives for a given event is extracted from the whole of the moment—in mind and world—passing as a flicker of time in an arbitrary point on a remote planet in one of many galaxies or universes.

Truth is commonly discussed as a direct relation of proposition to fact, but facts and propositions are outcomes of a cognitive process to which the individual has limited access. A determination of truth not only depends on the diachronic relations of a micro-temporal history but on the synchronic relations of attending to several objects at the same time; minimally, a statement, a perception and a judgment. This requires a series of mind/brain states, each arising in a distinct context, and each culminating in a different object. For a present occasion, the direct testimony of the individual gives greater weight to the subjective side of the equation. Statements about the future are predictions that might become true but are contingent on other events, known and unknown, or events foreseen that do not come to pass.

A pragmatic view of truth such as that of Peirce, James or Dewey, in which the practical consequences of a belief are determinants of its truth, or help to clarify its meaning, has an evolutionary character, in that truth is occasioned by the fitness of a belief to external affairs. A truth that rests on the elimination of the unfit (untrue) is what remains after errors are cleared away. The statement, *it is raining*, must be non-illusory, motivated by a desire to communicate a truthful observation, or not to deceive, and has to be set in an immediate spatiotemporal framework. Reality is the measure of truth but for some pragmatists the object of a truth is what we know of reality, i.e. the real is not an arbiter; it is established by truth.

In pragmatism, coherence and correspondence come together in verifiability, which allocates to truth that which can be validated, and thus conforms to what is external and real. The Cartesian emphasis on doubt applies to pragmatism in that the search for truth entails a rejection of what is false. While there are many variations on this approach, the elimination of doubt coincides with the actual process of thinking, in that the momentary genesis of a thought or perception involves a trimming of maladaptive possibilities as a configuration passes from category to instance or from generality to specificity. When all possibilities save one have been exhausted, what remains must be the truth. In this respect, the *via negativa* in thought is equivalent to the parsing of possible or actual error, i.e. maladaptive form or redundancy.

An instrumental role of thought as trial action entails a "best fit" to the situation or a correspondence of thought to reality, not a match of inner to outer but a *progression from inner to outer* in which the external is generated by the internal. Truth is what works in mind and world. A thought or act is true so long as the model of the world to which it belongs is adaptive. For coherence accounts, the theory decides what is true. For correspondence accounts, it is the facts that decide, even if facts are isolates within concepts. In correspondence, the match is presumed to engage final objects—propositions, worlds—not the subjective aim of thought to a proposition or world: for coherence, what matters are the contextual demands at every phase in transition.

Clinical study affirms there are no "rock-bottom" beliefs that are immune to alteration. It is the case that foundational beliefs are fairly resistant except in severe disorders of brain or psyche, e.g. delusion, hallucination, loss of personal identity, but it may be less the belief that has changed than constraints on its derivation to thought and action. This possibility is supported by the intermittent nature of false belief and the recurrence of beliefs that are true during intervals of lucidity, implying that the true belief "is there" but not implemented in behavior. Implicit *beliefs* are the source-concepts of objects and thoughts; implicit *values* are categorical primes for the derivation of feeling into action. Every act displays a value; every thought displays a belief. The truth of belief is constantly tested in behavior.

Belief adapts to truth in relation to experience. Value accompanies action and contributes to the nature and timing of acts. Fact is the final stage in value; truth is the final stage in fact. Every fact begins as a concept and terminates as a value. The beliefs and values that guide the process of fact-creation are implicit in every act or utterance, and exemplified in conscious thought and language. From this standpoint, truth is applicable to an occasion of experience, brief or long, but not, in my view, eternal; it is what fits best and endures (recurs) until a better fit comes along.

A scientific truth is, *ab origo*, not a creation but a discovery—an *uncovering* of what was previously not known—after which, once validated and accepted, it becomes part of habitual thinking against which other "truths" are judged. In science, truths require affirmation by argument and evidence and must survive attempts at refutation. While the tendency in science is confirmation, the argument of Popper that truth should be failed refutation is consistent with the microgenetic account of thinking. The scientific method of stating an hypothesis as a provisional truth and then attempting to prove or disprove it has some applicability to everyday trial and error behavior. Nietzsche's remark that truths are irrefutable errors is a skeptical account of this provisional status, yet leaves no room for disconfirmation.

On the microgenetic view, belief is an unconscious disposition to a certain line of thought and action. Beliefs are derived to thoughts or ideas as pre-perceptual images or preliminary speech-acts. Some implicit beliefs, such as those concerning subject–object relations, self-preservation, personal existence and, for most people, the reality of the external world are foundational in the derivation of conscious statements. Perhaps because these presuppositions are uniform across mental contents they can be ignored, but the effect of this strategy is to eliminate the path from belief to statement and the psychic targeting of statements as subjective aims. Core beliefs are general concepts of wide scope, an aspect of which is specified in a proposition. The propositional content is derived from the potential of the belief. One cannot take the proposition as a constituent of the belief, for the trajectory from belief to proposition is a complex, staged process, not a mere identification.

Microgenetic Concept of Truth

Truth is embedded in a relational context at successive points in its realization. Indeed, there is a kind of truth at each phase, though expressed in a non-rational vocabulary, e.g. symbolism, analogy, as in poetry, dream and animistic thought. A fact is embedded in a world. When it is thought or felt or pronounced, fact issues from a relational matrix as an artificial isolate, but the context of every thought or utterance is a hierarchy of categories that begins and ends with self and world. A statement, and the world it refers to, grow out of imagery. The image-substrate that gives rise to statements is equivalent to that which gives rise to world events. In what sense is the imaginal basis of perception, or the verbal imagery anticipating speech, less "truthful" than the final commitment? A rational proposition is close to an object in the world, but truth is established in the relation over levels in world creation.

There are many forms of truth apart from isolated propositions connected to local facts. There is truth that expresses genuine feeling or reveals integrity and authenticity. The employment of propositions that are true or false conveys meaning as reference but does not touch the deeper meanings of which the statement is a surface mark. Logic and science constitute a family of isolated statements that combine to what passes for context; a topic, a field of inquiry, in which each particular has its proper share. Water is H_2O, boils at 100°C and is about 70% of the human body. These truths or like-statements share a family resemblance, but their import to feeling or meaning, e.g. thirst, cold, is less amenable to a propositional formulation. This is another way of saying that a proposition avoids the deeper truth that shines as from below. Except for the precision of mathematical logic, one reason that truth is so elusive is that belief is disposition until it is frozen in fact, and much of the potential in thought is spent the moment a sentence is evoked. The attempt to capture a truth buried in highly constrained sentences gives rise to alternate modes of language and imagery, such as allusion and allegory. Some languages, such as Chinese ideograms, strive for meaning by incorporating different properties or analogous meanings of a word from several directions. This is a perspectival approach to reference instead of a direct assault. Roundabout ways of truth-finding and sampling context prior to exact formulation express phases in sentence-production antecedent to the final utterance.

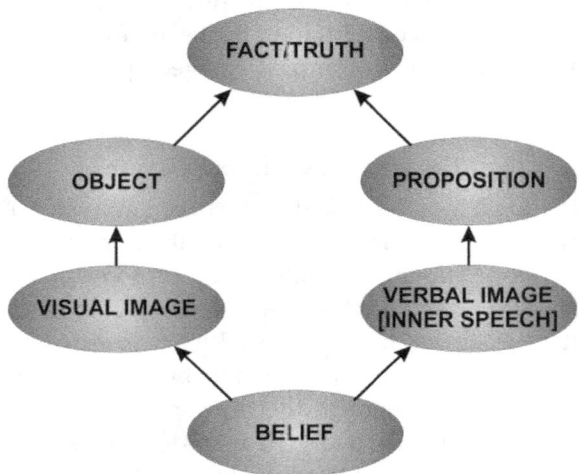

Fig. 1: Implicit or unconscious beliefs give rise, *inter alia*, to visual and verbal images. Verbal imagery is inner speech. Constraints applied to the developing images transform them to objects. The image does not become an object; rather, the ground of imagery transforms to perception. An object can be a visual object, a fact in the world, a mental statement of utterance (see text below).

The appeal is not to alternate or primitive modes of thought but to phases in thought-development that prefigure propositions and truth-judgments. These are not merely different types of truth, such as poetry and science. The poetic or symbolic represents the context out of which true statements arise. A proposition is a final phase in the specification of (the ground of) verbal thought. An object or fact is a final phase in the specification of (the ground of) visual imagery. A fact that evolves out of perception is an instance in the field that has propositional support, while the proposition that evolves out of the potential of thought is supported by the externalized object. Fact, object and utterance are equivalent phases in the momentary derivation of an act of cognition. Truth is mapped across these finalities, but its ground is the process out of which they develop (Fig. 1).

Reality is known or inferred from objects and factual truths, but a more precise formulation is that reality shapes imagery to model the physical world. The perceptual derivation of this process — an object or event — becomes a true fact when it maps to the linguistic derivation — a rational or factual proposition. The mapping across these derivatives of (the substrates of) imagery is termed a truth. We also apply truth to the correlation of feeling with language and perception, as in the emotions aroused by fine music, though such truths are personal and subjective, and not readily validated by others. We say that events have the "ring of truth", or feel right, or touch us deeply, and for some the deepest truths are ineffable. These examples are, of course, outside the usual understanding of truth, much as hallucination is outside that of perception, but they do provide a glimpse of the infrastructure on which judgments and agreements as to truth depend.

From such observations, we learn that the real, whatever it may be, is not determined by truth, and that truth is not a correspondence of a statement to the real. The apparent real is the correlation across endpoints in cognition, while the actual real is outside thought. The real is what constrains perception to a model — in words and objects — of what reality must be like. The model is constantly tested against reality and revised to achieve optimal success. With respect to truth, success can be defined as thought or action that is adaptive, in the evolutionary sense, to the world of perception in that it helps to ensure the survival of individual and society, and the reproduction (recurrence) of mind/brain states. However, the correlation from which reality or truth is inferred is the concordance of language and perceptual actualities in mind, not of language in the mind and events in the world. This correlation is a kind of consensus across cognitive modalities in the waking state similar to that which accounts for the reality of events in the dream state. If the verbal imagery of dream could be framed as a proposition, or if decisions and judgments could be made in dream, we would claim the truth of a "correspondence". The analogy with dream images and the feeling of the reality of dream by way of a collusion of the perceptual modalities reinforces the claim that the linguistic outcome of thought is primarily perceptual so there is, ultimately, an intra-psychic correlation across two different modes of perceptual imagery.

The conventional theory of correspondence is thus revised to a correlation between perceptual endpoints that can be tested in action and recurrent perception. Reality is not the complement of perception with rational propositions but an inference from this correlation that is provisional and based on successful adaptations. The perceptual correlates of a truth judgment are the residuals of a process of elimination in which antecedent context is inapparent because it is further specified. The final content is obligated by extrinsic constraints, but also by the coherence with its formative history, a context implicit in objects and statements but relegated to the background. All phases leading to a proposition or object, from their onset in belief, their passage through symbolic or animistic thinking, to the substrates of imagery close to the perceptual surface, are simultaneous and thus are ingredient in whatever actualizes. The conventional account of coherence can be interpreted from a process standpoint as the conformance of whole–part relations over a series of phases, such that the final content is consistent with the source-beliefs out of which it develops.

Theoretical Note Three

On the Direction of Time

> *Our fixed direction (in time) is given solely by the advent of new arrivals.*
> —Bradley (1893, p. 89)

The aim of this note is to examine the truth of Bradley's remark. Life advances into novelty but is novelty responsible for the direction of the advance? What determines novelty? Does it depend on the consciousness of change or the recognition of difference? In other words, does it depend on human cognition or is it universal in nature built into the process of change? Is the individuality of subject and perspective ingredient in world process? If one assumes causal progression, the entirety of the universe would undergo novel change each moment. What magnitude of identity does it take to offset novelty? If an event is 99% recurrent and 1% novel, is this sufficient for a novel occasion? One might suppose that by new arrivals Bradley means the totality of personal experience, not the newness of a particular event. If an event is identical to a preceding event, is the person, or universe, identical across events whether the events occur at the same or different times? Since two objects cannot occupy the same space at the same time, and entail different relations to an observer, an absolute identity of observables is not possible. If novelty determines time-direction, what degree of novelty is required for forward motion? In all of this, what is identity and what are the implications of sameness?

The literature on identity beginning with Leibniz implies a judgment as to the properties of substances as mental objects, or of substances as mind-independent entities. In process ontology, the actualization that deposits a thing—an object or entity—is ingredient in what the thing is. Common sense would have it that two new and otherwise identical combs are the same, so identity at a theoretical level is quite different from that in everyday language. Nor would we be inclined to say, even theoretically, that one of the two combs is novel in relation to the other by virtue of theoretical distinctions. Ordinarily, novelty implies more than difference; it suggests genuine newness. The words in a poem of Keats are no different from those in a dictionary.

The sounds of the words and their organization are novel, though this would be true of a random arrangement. What is genuinely new is the feeling that is evoked, the beauty and the meaning, and these are subjective features.

Advance and Retreat

Leaving aside for the moment these difficulties and taking at face value Bradley's statement, the implication is that novelty drives time-direction. However, why should forward direction be driven by novelty any more than the reverse since a movement to the past would be novel to the observer if it were not personally experienced or erased from memory? If time-direction should begin with the death of the universe and progress to its birth, there would still be a sequence of novel encounters. If novelty refers to the totality of experience, a shedding of acquisitions would give a novel state. A present that becomes past — like a past that becomes present — entails novelty even in retracing prior occasions. The physical universe a moment ago differs from that of the present. Any motion or change is novel in relation to its antecedents. This comes down to the question, is novelty an attribute of change in physical nature independent of mind, or does it require a comparison to prior events? Change that occurs in the reverse direction assumes that the forward direction has already been traversed. For this, a judgment of past and present is required. If direction is due to novelty in mind, recurrence would alter forward progression and change would move to the past or the future depending on the sameness or difference of each occasion.

This implies pre-existing order, since without prior direction reversal is meaningless. The process of change may be identical from one realization to another, but that which is changing, e.g. the realization of an entity or mental state, differs from one instantiation to the next, since a novel entity, for the next cycle of recurrence, incorporates the process through which it actualizes. On the one hand, in the evolution of life forms, consciousness of new arrivals as a source of time-direction is the correlate of obvious anisotropy in nature, e.g. evolution, growth, decay. On the other hand, nature is cyclical, e.g. on a geological, seasonal, lunar, circadian or momentary scale. Time is linear, with forward movement in cyclical process. The cycle of sleep and wakefulness that recurs throughout life is compatible with growing older every day. Even if cycles are not precisely identical, and even with a grain of novelty in a field of repetition, this does not impact the sense of fixed direction, though recurrence must be assimilated to anisotropy as a mirror of natural process.

A return to preceding states is not possible in the absence of a forward direction up to the point of reversal. Unlike a film in reverse where recurrence is for one object, what is required is a reversal or comprehensive arrest of change. A reversal of one time series does not include a reversal of all, i.e. each re-encounter would be enfolded in some forward change. Moreover, if the sequence from past to present has already occurred, an alteration of direction regardless of novelty can only be a reversal. For this reason, a direction from present to past presupposes a reversal of forward advance.

With either novelty or recurrence, or some mix of the two, direction in time in an individual mind might be expected to reflect the dominant mode of experience, e.g. memory or action. However, if only a trace of novelty is necessary for anisotropic time, i.e. if any change is a total change, *degree* of novelty is irrelevant to forward motion. A loss of a single skin cell, one hair out of place, is a change in the total organism. Otherwise, the massive recurrence (and minimal novelty) in every event or act of cognition, combined with the cyclical character of process in animate and inanimate nature, would confound time-direction or give the experience—which, in fact, is the true nature of subjective time—of the replication of a motionless point (below). Of interest, the one state in which the world of an individual appears fully novel without a past or future is dream, in which the motion of images is, I would argue, ordered on waking but simultaneous in the dream.[1]

Novelty and Memory

The distinction of novel experience and memorial content—thought/ perception and recollection—is central to Bradley's argument. A thought that is identical to a previous thought still moves forward or occurs in the context of earlier and later. A repetitious idea does not affect the passage of time. How do we know a thought or perception is past or present—a new arrival or a repeat visit—unless it is compared to past experience or activates identical neural configurations? It is possible but unlikely that an exact comparison is implicit in every mental state.[2] Memory is never exactly true to perception. The problem

---

[1] In meditation, in deep concentration and absorbing thought, or listening intently to music, time may seem to stand still though the images and sounds are novel or, in familiar music, if not novel, at least show an advance as the surrounding world fades into the background, e.g. Eliot's "you are the music while the music lasts." Music is a good example of a sequence that is fixed, and plays out in a known and predictable manner, yet time continues to advance.

[2] I am sure many of us have had the experience of reading an essay or seeing a film for the second time after a long interval and not recalling the original

is that the brain changes each moment as does recall so it is unlikely a match could occur.[3] Every event—novel or recurrent—is experienced by a novel individual. The object or experience is not isolable from the internal state of the experiencing organism. Except in a Laplacean universe, novelty is not in the arrivals but in the organism that is changing every moment. Memory is not coded as past but is accompanied by a feeling of pastness due to an attenuated actualization, i.e. a pre-perception. The fluid relation of thought and memory and the continuum of recall with perception reinforce this interpretation and help to account for phenomena such as *déjà vu*.

Novelty in thought depends on recurrence and forgetting. Cases of prodigious memory are often deficient in abstract thought;[4] an exuberance of memories forecloses new ideas (Luria, 1968). The irrelevant must be suppressed to elicit the salient. If the thinker is unaware that an experience or idea is recurrent, does this negate its novelty? Were someone to forget that he uses the same comb every day, to what extent is using the comb novel? Is every occasion novel in spite of constant recurrence? The newness of momentary circumstance is sufficient to guarantee novelty even if a thought or perception is repeated, i.e. the thinker is novel if not the thought. The question is how to characterize novelty in a "new arrival" if the content is recurrent but the organism and its *environment* have changed.

Deviation in memory is an occasion of thought. A fabrication or mistake in recall is a transition to thinking but it is not a novelty that explains time-direction. The knowledge and habit that go into thinking are occasions of memory ingredient in thought. If the memorial basis of thought and perception is recognized, what would it take to compromise forward momentum or for an individual to be frozen in time? A person with novelty in every thought would still not be propelled to the future, partly because memory, not novelty, underlies time and partly because prior beliefs and values, language skills and concepts are the foundation of the mental state, the major part of which, including the surge to a future, involves past events in relation to a

---

except for some fragments. A personal example is that on re-reading Bergson's *Creative Evolution* at the age of 50, to my surprise I came upon passages marked off on a first reading when I was 16 that were directly relevant to ideas that did not arise for another 15 years. Are neural systems activated on the initial exposure entrained again but without the correlated memory? What accounts for the recognition of some past events, the forgetting of others, the sketchy recall of still others and the uncertainty as to whether a thought is novel or recurrent?

[3] Cf. picture-script memory (McCullough, 1965).
[4] A literary example is the story *Funes the Memorius* by Borges.

changing present. When explicit recall is deficient, as in amnesia with a contraction of duration-estimation, time-experience is not static nor does it move to the future; rather, in extreme cases, the self is experienced as isolated in time.

If one imagines a state in which there is only memory without perceptual experience, such as the survival of mind up to the instant of death, as chillingly described by Philip Roth (2008), the present would contain only memories, presumably without forward movement. The intuition is that memories—similar or changing slightly in each recurrence—would be revived into the present without backward movement in time. If new arrivals give advance in time, memories do not give retreat. A continuous revival of the past would not lead to time-reversal but to an arrest of forward movement. One supposes that unconscious memories could surface as novel recurrences. Bradley might distinguish psyche and object, though his philosophy mandates a continuum, such that repeated perception of the same world would disrupt forward direction. On the other hand, for time to move backward requires, on this interpretation, a precise unraveling of the onion skin of experience.

If time-direction is determined by a past to present series irrespective of perceptual or memorial content, the direction could establish novelty even if the content is recurrent. On this view, novelty is unrelated to direction but is extracted from dissimilarities in momentary experience. Regardless of whether present content is identical to that in the past, similarity and novelty are judgments, not temporal determinants. Conscious novelty hinges on this judgment. In comparing an event to a prior event, even if the event is ostensibly the same, it is still novel, as an addition to the sum of experience.

Actualization and Subjective Aim

I have argued that irreversibility in primitive life forms develops out of isotropic energy in inanimate matter (Chapter 5). Directionality arises in the surge of the subjective (intrinsic) aim. This evolves from simple tropisms to purposeful action in animals, then to intentionality in the sphere of concepts. In the progression from primitive organism to human cognition, the expansion of trends in anisotropic feeling and the implication of panpsychism relate to the before/after of time-direction as a mind-independent outcome of actualization.

The inference to be drawn is that, from the standpoint of momentary occurrence, subjective time does not move in any direction. Rather, the serial unfolding of events, considered apart from historical change, aging and the observed sequence of life-events, is layered in the simultaneity of the mental state. The state deposits a present and gives the sense of antecedents leading to it. *The subjective aim of feeling, and the actualization of the mental state from past to present, are responsible for the forward direction.* The embedding of prior occasions in the immediacy of the present, that is, in the iteration of actual objects, gives novel occurrences that reinforce the illusion of advance, which in fact is due to an incompleteness of former occasions irrespective of novelty. However, a caveat is in order in that a world that does not change, or one in which every replicate is identical, would probably, and fairly quickly, lead to confusion, as in snow-blindness with a loss of objects, or to disorientation in space and, possibly, in time in sensory deprivation.

The claim is that time order, direction and subjective aim resolve to the process through which aim and order are established in the mental state. Take the common example of listening to music. In order to hear a melody it is not sufficient to say that past tones are remembered. The past tones are revived in the order of their occurrence and heard in an illusory present that spans—actually, compresses in a single mental state—successive segments of sound. The present is generated in the relation over phases within a mental state, in which the phases correspond to revived prior presents; e.g. when successive tones at T-1, T-2 and T-3 are all revived in the tones at T-4. The tones are "stacked" within the state—a simultaneous epoch—with the sequence actualizing in the duration of the specious present. The perceptual endpoint of the state deposits the passing present while the overlap of states gives the experience of a rolling transition in time.

Hume wrote that one cannot catch a perception. This is because it is constantly being replaced, along with the stacked series enfolded in the present state. The realization of a perception is the satisfaction of a subjective aim over a corpus of immediately prior but incompletely revived memorial phases embedded as incomplete objects. The trailing sequence of tones allows the melody to be perceived in a direction from past (memorial) to present (perceptual) in time-consciousness.

Change and Stability

Within the overlap, mental states can be similar or dissimilar. An orchestra is relatively similar from one moment to the next, and appears as a stable or relatively solid object, while music differs across states and is perceived as dynamic and changing. If the difference is too slight or unnoticed, as in a piece by Philip Glass, a Sufi prayer, a drumbeat or a Buddhist chant, the effect may be hypnotic or one of boredom. If the differences in tones are too great, the music will be perceived as incoherent. Sameness or difference is perceived within a larger framework of phrases or patterns that may also show novelty or recurrence. A repeat in a Schubert quartet may be a delight to the listener, especially if the music is familiar and the repeat is anticipated, while in other compositions this can be wearisome. A rhythm that is repeated in rapid succession can arouse the listener to dance. The common refrains in song or poetry are often much enjoyed. The effect of sameness and difference has to be interpreted in relation to patterns within which the recurrence is enfolded. In contrast, identity and novelty are metaphysical problems that are independent of individual subjectivity. Still, it may be that an understanding of attributes of the mental state can shed light on phenomena in material nature.

The relation of simultaneity to order still holds with respect to material nature. In human cognition, we interpret novelty as what is genuinely new or creative; in physical nature, novelty is the absence of identity with respect to some property. There is the question of identity for elementary particles, and that for combinations. As to the former, novelty lies not only in spatial location and/or velocity but in the intrinsic process through which the entity recurs. At this level, novelty is minimal and highly theoretical. The situation differs in aggregates.

Thus, while one molecule of water seems much the same as another, a molecule in relation to many others gives liquidity. In water, the removal of one molecule has no noticeable effect, but if this continues, the residual ceases to be liquid. One can say that a defining feature of a molecule of water is the potential for liquidity or that the properties of a single molecule include the potential for liquidity in a multitude. At some point we can say something novel has occurred apart from the addition. The same problem occurs with ice and vapor, and applies to a gas that becomes explosive at a certain heat. We say, some threshold has been reached, but each stage in the progression to this threshold alters the possibility of occurrence, which means that each additional molecule or change in velocity changes the potential of every constituent.

Properties are like attachments that can be added or removed according to conditions. They are also theoretical objects in the human mind. If primary properties such as mass are as subjective as secondary ones, mentality is part of what an object is. Process theory takes an internalist perspective in which the coming-into-existence of an entity is decisive in determining what the entity is. The process of becoming corresponds to one cycle of completion as a precursor of the subjective aim, with the internal relations in the process through which objects are generated providing the basis of change.

In contrast to isotropic energy in physical entities the becoming of feeling in living things is unidirectional. The transition to an outcome — the subjective aim — is the seed of time-direction. Actualization in an epoch goes from before to after or from earlier to later. This sequence also characterizes the progression in a mental state. In human cognition, a present is created between earlier as past, and later as future. Earlier phases correspond with memory, while the endpoint corresponds with the perceptual present. The transposition of antecedents in the mental state to a longitudinal series leading from past to present gives the forward direction in time, while the surge to actuality and incessant replacement give motion to an oncoming future.

Theoretical Note Four
Origins of Religious Belief

Introduction

One can think of the world's religions as dynamic systems that evolve over generations in which features of earlier religions are carried into later ones. One scholar put it that the aim of the study of religions is to grasp the religions of the world as manifestations of the world's religion. That is, a set of critical features are preserved, modified and shared by all religions, modern and ancient. What are these features and why is the belief in a higher power, a supra-individual mind or a divine presence so widespread when the evidence to support it is so meager?

Religious belief must trace to essential features of the human mind. Some have postulated that religious belief is innate. Why the need, why the attributes, why the worship and unwavering faith? To say that faith mitigates fear of death or alleviates suffering or, as Freud argued, as recorded in the Epistles of St. John, expresses and/or satisfies the need for an authoritative father—true or not—does not explain the attributes of god, the concept of soul and spirit, devotion and appeasement. The need for an all-powerful father might as well derive from the pack instinct of social animals. In many early religions, heaven or the sun as father and earth or the moon as mother are common themes. The enchantment of nature, the need to account for a catastrophe, for illness and for death, does not lead inevitably to explanations on a religious basis. Need is fundamental in all organisms but in humans it accompanies the worship of a higher power who can punish and protect.

Qualities that justify belief, such as perfection, all-knowing, eternal, punitive and merciful, or behaviors such as awe and submission, are ingredient in conviction, while the accoutrements of religion—rites and dogmas—instill and strengthen devotion and mediate an encounter with the unknowable. These are necessary since direct contemplation of the eternal seems as implausible as an immediate awareness of nothingness.

This note attempts to fill the gap between faith and justification by demonstrating that religious thought and related concepts of mystical awareness owe to patterns of realization in the mind/brain state. It is a given that the structure of the state is laid down by morphogenetic process in concert with genetic determinants and the animal inheritance, and reinforced by sensibility so that endogenous process adapts to nature and the social environment. *The central features of microgenetic theory that pertain to religious thought are the derivation of thought and perception over a hierarchy of subjective categories, the relation of being as category to becoming as process in the arising of the mental state, the passage from unconscious simultaneity to conscious temporality in subjective time experience, the continuity from mind to world, the transition from feeling to value, and the whole-part transition.* These features of human mind compel a belief in a higher being.

If process in the mental state underlies the concept of god, and if the foundations of the state determine the attributes of a deity, one could suppose that the relevant properties of the human mind develop over an evolutionary process that prefigures the higher mentality. In my view and for Whitehead as well, random variation, fitness and contingency are incomplete explanations of evolutionary advance. There is a direction to actuality and subjective aim. If faith is conditioned on features of the mental state, proto-psychic qualities in primitive life forms might lead the evolution of human mind to an inevitable intuition of god. The panpsychic alternative to the origin of human mentality rests on the unidirectionality or teleology of evolutionary process, the categorical nature of being, and process or becoming as a recurrent exemplification of feeling.

On these grounds, one could suppose the human mind/brain evolved in such a way as to satisfy its creator. This is not, however, an argument for god's existence but, rather, a theory of how the belief in god comes about. The uniformity of process underlying a becoming-into-being and the progression over the evolutionary series eventuates in patterns of subjectivity that are sources of religious thought. The goal is to collapse a diversity of beliefs to shared attributes as a framework for understanding the universality of religion and the strength of conviction in an age of technical advance, scientific materialism and skepticism in which a machine nature and a computer brain leave little room for faith. The scientist looks at causal laws that operate outside the influence of god or mind-eternal, with object-features inserted in the mind/brain as logical solids. In contrast, for microgenesis, objects are specifications of categories and mind is an organism with a capacity for novelty and adaptation.

Religion conceives a mentality in physical nature from which all things arise, or a mentality outside nature to which all things respond. Yet, the affirmation or denial of the truth of religious belief is not settled in a dispute with nature. The nature we know is that of experience, so it is not on earth or in heaven that the origins of deity and the impetus to religious thought are to be found, but in patterns of personal mentality that, for religious thinking, are its authentic home. In the spirit of Whitehead's remark, that "religion is what the individual does with his solitariness", I refer here to the nature of mind, the individual quest and the encounter with deity, in line with the teachings of Buddha or Christ, not the tradition of the lamas or the church of St. Paul. Religious thought, though seemingly projected onto nature, has its object within the mind, not external to it, neither in science nor cosmology, and not solely because nature is, if it is, a conception or that a causal nature is anterior to mind, but that religious thought, feeling and spiritual need are created and justified by the pattern of mentality.

While the focus is not on the details of practice, some features of the mental state can be seen in religious dogma. An example is the interpretation by Jakob Boehme of the trinity as a passage from God as potential to Christ as actual with the Holy Ghost the process of transition. The becoming-into-being is consistent with microgenetic theory. Similarly, in mystical Judaism the birth of the universe is believed to pass from the mind of God to a not-nothing that takes on direction. This belief also resonates with microgenesis, and is reminiscent of findings by the Würzburg School in psychology of the *Bewusstseinslage* and *Bewusstheit* as stages in the birth of a single thought.

The Sources of Religious Belief

Through a series of partitions, the mind/brain state deposits rational thought near an analytic endpoint, with object-perception the final phase. For surface cognition, the particulars of the world seem independent and unhinged from a guiding spirit, especially from the mind that perceives them. Logical statements or mathematical propositions detach as topics of consensual study. Since the unconscious antecedents of the mental state are distant from the manifold of surface realizations, the individual is not conscious of the continuity to substance-like outcomes. A conscious individual centered in mind-external is oblivious to ancestral form. In the actualization process, feeling as value accompanies the object-formation and actualizes as fact. The transition from belief to fact is filtered through phases of progressive adaptation. The more successful the adaptation, the closer thought and proposition are to objectivity. An object-centered perspective exchanges the process of thinking for an attachment to the products of thought. The mystic takes the opposite path, forsaking objects and settled fact for an inward retreat to layers of mind beneath the encrustations of adaptation and habit. The sources of religious belief are uncovered in a descent from the diversity of surface-mind to the deeper uniformity of mind-internal.

The hierarchic structure of the mental state underlies the transition from depth to surface and the reverse. The ascent to what lies beyond and the descent to what lies within suppose a theory of mind that entails a hierarchic succession of phases. One direction is a progression to higher levels from subjective to objective, from self to world or past to present. This is the foundation of the belief in a higher or noetic sphere of supra-individual mentality. The opposing direction is a withdrawal to the core of individual mind to an immediate apprehension of oneness with god. One could say that an object loses its absoluteness by descending into thought and becoming conditioned by the consciousness that contemplates it. Subjectivity, immediacy and proximity to the sources of belief give the conviction that is lacking in object knowledge. Certainty in object knowledge is provisional, while conviction in subjective belief is unshakeable.

Inference of Deity from Cognitive Process

More precisely, the inward retreat from objects to images through fields of meaning-relations to the core of the mental state gives an intuition of foundational categories that prefigure and enfold the individual series. These categorical underpinnings tender the idea of a universal category — an Absolute or eternal consciousness — with all entities conceived as particulars in the mind of god. The progression from unconscious simultaneity to conscious temporal order — the becoming-into-being — is the relation of the eternal to the momentary. The temporal quality of lived experience in relation to a non-temporal unconscious gives the experience of an individual life in relation to mind-eternal. The eternal of god's mind is extrapolated from unconscious category, thus the relation of religious belief to dream and myth. The intuition of organic strata in individual-mind is the immediate experience of the mystic in a withdrawal through layered unconscious fields to a source beneath words. The centrality of feeling is a source of unity, conviction and the bond with god or nature. It is also an internalizing force that brings this relation firmly into the intra-psychic sphere. While knowledge is directed to an object, feeling brings the immediacy of the internal to the encounter with god. Finally, the relinquishment of self in an indivisible field of feeling as valuation becomes an immersion in oneness with the living spirit.

The cascade of categories, and the derivation of wholes to parts, leads from simultaneity to serial order — from the timeless to the temporal — in a passage from an eternal past to a momentary present that translates to a finite point in a surround of infinite space and time. The merging of the finite with the infinite, and the evanescent with the eternal, are essential themes in the relation of an individual to the universal. The past, the present and a presentiment of the future concur in the illusory now. This illusory or *specious* present is also a conception in the religious imagination, for example in a world as the dream of a sleeping Brahma. The goal of meditation is an expansion of the now to the all-embracing present of god's mind. Meister Eckart wrote "the aim of man is beyond the temporal — in the serene region of the everlasting present"; Coomaraswandy (1993ed) wrote in a similar way that in meditation, "contemplation on time [is] directed outwards to the immediate realization of ever greater and greater durations and pursued until the whole of time can be experienced now."

Another feature of mind that is central to religion is animism, in which the world is continuous with inner-mind. Animism is not merely a primitive stage in the history of religious thought, but is an active phase in the mental state in which nature is alive with psychic entities. The feeling that accompanies the object-development gives a valuation of the external in a non-material world. Concealed remnants of animism are uncovered in mature religious feeling. For example, the belief in an after-life is probably derived from the belief in a transcendent world of spirit. Belief and ritual propagate by metaphor and symbolism. The passage from animal gods and totemic objects to anthropomorphism is a shift of spirit from external objects to properties of the self. Inner-mind exports subjectivity to nature in value, purpose and meaning that become the all-knowing, all powerful or all-merciful of an eternal father, or a guiding spirit or vital force.

The iterated passage over phases from core to surface or from self to world leads from uniformity to multiplicity. Unity at the base of individual mind is obscured by the outward trend to serial order and the manifold of external space. The derivation of particulars from categories is a microcosm of the origination of nature in the mind of god. One can say the all-in-one gives the one-in-all through successive divisions. Invisible nature at the core of mind, the unseen uniformity of natural process, gives the many at the surface as actualizations of the One at the depths. The self that arises at the conscious floor of the mental state gives birth to the world as a singular image that distributes into a manifold of objects and their valuations.

Reason is a way station in the realization of diversity. It delimits possibility and is a pale copy of the wholeness lost in division. The rational tends to be constrained and predictable, while creative advance begins with the indefinite when it *uncovers* latent concepts. The idea of thought as an uncovering recalls the Socratic notion that teaching draws out what one implicitly knows. Heidegger used the term *uncovering* in relation to truth.[1] On the conventional view, reason fractures cognition into elements that can be studied in isolation. But isolates that arise as contrasts or oppositions take us further from unity. The greater the analysis the more piecemeal the construct and the more distant from the wholeness to which it aspires. An object of thought is part of the thought it attends to. Thought faces inward to the self and outward to its aims, yet the mental state is an epochal whole in which self and other arise out of a common ground. For Buber, beneath the *I* and *Thou* is a unity of both perspectives. The descent that uncovers the source of the other is the same inward journey that reveals the deep subjectivity and shared feeling that are the experience of oneness with god and the basis of love, given and received.

Faith is belief in opposition to objectivity; unlike reason, it is not tethered to fact. The absoluteness of faith conflicts with the provisional of reason. For Kierkegaard, "faith is the contradiction between the infinite passion of inwardness and the objective uncertainty." One could say, the conviction of subjective belief is refractory to objective certainty. The provisional in science vitiates certainty and accentuates the impact of revelation and the unshakeable conviction of subjective knowing.

A residue of nature-mysticism traversed in every mental state compels a form of panpsychism as a matter of faith. In the concept of Indra's net, the Buddha-nature extends to the smallest particles; god's creation is everywhere filled with spirit. The distribution of mind into the smallest entities is reversed in the evolution of the proto-psychic to human mind. Whether mind passes into entities or evolves from them, to fully understand one thing means to understand all things — Blake's universe in a grain of sand, Tennyson's flower.[2] In the fractionation of mind to the smallest objects, and the evolution of basic entities to consciousness, mind-external is not projected on objects but individuates with them from an archetypal ground. The consolation of individuality as transient appearance in mind-external is life as a finite mode in the mind of god.

Feeling goes outward with the object-formation and gives object-value by way of progressive delimitation. Every individuation forecloses alternatives. A path chosen leaves a world of unactualized possibilities. Eckart wrote that "only the hand that erases can see the true

---

[1] Heidegger's terms were *aletheia* or unconcealment, and *Entdeckendheit* or uncoveredness.

[2] "Little flower — but if I could understand
What you are, root and all, and all in all,
I should know what God and man is."

thing." Evolutionary pressures eliminate maladaptive possibilities. Sensation carves out objects. Every phase in mind, from unconscious drive to conscious action, is a compromise in finality. Mind is an evanescent set of fleeting categories that partition to actualities as negative images of the real. The germ of this process is found in the most rudimentary life forms. What is the difference between a flower that bends to the sun and a bee that alights on the flower? Each thing moves in some direction over other possibilities. Adaptation is one reason for this though all choices are not instrumental and the will to survive is a value apart from the means to insure it. These implicit tendencies are distant forerunners of value and conscious choice.

An object is a thought parsed by sensation to fit changing patterns of an unknowable reality. Object- and value-formation transport concepts and feeling—being and becoming—into an outer world of perception. The relative worth of things perceived, with love the ultimate in positive valuation, is intuitively transposed to the infinite value of god. The creation of value in mental objects becomes god's valuation of a world of his creation. The compassion for others that is the summit of personal valuation becomes god's love for all living things. Conversely, fear as a negative polarity in value-formation becomes a shuddering and foreboding at the limitless power of god, while the development to conscious thought and desire through the refinements of conceptual-feeling continues in religious thought as an aim to the hereafter through the worth of individual acts. In many individuals, an extreme event, a vulnerability of self, a recognition of a life imperiled and a limitation on the possibilities of action condition a turn to faith. The fractionation of the mental state in adaptation to the world reinforces the sense of limitation, while a suspension of ego accompanies a receptivity in which the partiality of thought obtains completion in the perfect wholeness of god's mind.

References

Al-Azm, S. (1967) *Kant's Theory of Time*, New York: Philosophical Library.
Atmanspacher, H. (2012) Dual-aspect monism à la Pauli and Jung, *Journal of Consciousness Studies*, 19, pp. 96–120.
Atmanspacher, H. & Filk, T. (2011) Contra classical causality, *Journal of Consciousness Studies*, 19, pp. 95–116.
Bachmann, T. (2000) *Microgenetic Approach to the Conscious Mind,*. Amsterdam: John Benjamins.
Bachmann, T. (2006) Microgenesis of perception: Conceptual, psychophysical, and neurobiological aspects, in Ögmen, H. & Breitmeyer, B. (eds.) *The First Half Second: The Microgenesis and Temporal Dynamics of Unconscious and Conscious Visual Processes*, pp. 11–33, Cambridge, MA: MIT Press.
Bachmann, T. (2012) How does microgenetic theory square with evidence from cognitive neuroscience?, in Pachalska, M. & Kropotov, Y. (eds.) *Psychology, Neuropsychology and Neurophysiology: Studies in Microgenetic Theory*, in preparation.
Bacon, J., Campbell, K. & Reinhardt, L. (eds.) (1993) *Ontology, Causality and Mind*, Cambridge: Cambridge University Press.
Bain, A. (1868ed) *The Senses and the Intellect*, London: Longmans, Green, & Co.
Bartlett, F. (1958) *Thinking,*London: George Allen and Unwin.
Bender, M. & Krieger, H. (1951) Visual function in perimetrically blind fields, *Archives of Neurology and Psychiatry*, 65, pp. 72–99.
Bergson, H. (1896) *Matière et Mémoire*, Paris.
Bergson, H. (1910) *Time and Free Will*, English translation of: *Essai sur les données immédiates de la conscience*, London: Swan Sonnenschein & Co.
Bergson, H. (1920ed) *Creative Evolution*, Mitchell, A. (trans.), London: Macmillan.
Bergson, H. (1923) *Durée et simultanéité*, 2nd ed., Paris: Felix Alcan.
Bernstein, N. (1967) *The Coordination and Regulation of Movements*, London, Pergamon.

Bertalanffy, L. (1968ed) *General System Theory: Foundations, Development, Applications*, New York: Braziller.
Birch, C. (1990) Value and meaning of process thought, *Process Studies*, 19, p. 224.
Bosanquet, B. (1904) *Psychology of the Moral Self*, London: Macmillan.
Bradford, D.T. (2014) *The Spiritual Tradition in Eastern Christianity: Ascetic Psychology, Mystic Experience, and Physical Practices*, Leuven: Peeters Publishing.
Bradley, F. (1893) *Appearance and Reality*, Oxford: Oxford University Press.
Brentano, F. (1874/1973) *Psychology from an Empirical Standpoint*, London: Routledge and Kegan Paul.
Bricklin, J. (2014) *The Illusion of Will, Self and Time: William James's Reluctant Guide to Enlightenment*, in preparation.
Brown, J.W. (1983) Microstructure of perception, *Cognition and Brain Theory*, 6, pp. 145–184.
Brown, J.W. (1985) Clinical evidence for the concept of levels in action and perception, *J. Neurolinguistics*, 1, pp. 79–87.
Brown, J.W. (1987) Review of: *The Mind at Work and Play*, *Journal of Nervous and Mental Disease*, 175, pp. 311–312.
Brown, J.W. (1988) *Life of the Mind*, Hillsdale, NJ: Erlbaum.
Brown, J.W. (1991) *Self and Process*, New York: Springer-Verlag.
Brown, J.W. (1994) Morphogenesis and mental process, *Development and Psychopathology*, 6, pp. 551–563.
Brown, J.W. (1996) *Time, Will and Mental Process*, New York: Plenum.
Brown, J.W. (1999) Microgenesis and Buddhism: The concept of momentariness, *Philosophy East and West*, 49, pp. 261–277.
Brown, J.W. (2003) Value in mind and nature, in Riffert, F. & Weber, M. (eds.) *Searching for New Contrasts*, pp. 37–59, Frankfurt: Peter Lang.
Brown, J.W. (2004) The illusory and the real, *Mind and Matter*, 2, pp. 37–59.
Brown, J.W. (2005) *Process and the Authentic Life*, Frankfurt: Ontos Verlag.
Brown, J.W. (2005a) Genetic psychology and process philosophy, *Process Studies*, 34, pp. 33–44.
Brown, J.W. (2008) Consciousness: Some microgenetic principles, in Riffert, F. & Sander, H.-J. (eds.) *Researching with Whitehead: System and Adventure*, Munich: Alber Verlag.
Brown, J.W. (2009) Inner speech, *Aphasiology*, 23, pp. 531–543.
Brown, J.W. (2010) *Neuropsychological Foundations of Conscious Experience*, Belgium: Chromatika.
Brown, J.W. (2010a) Simultaneity and serial order, *Journal of Consciousness Studies*, 17, pp. 7–40.

Brown, J.W. (2011) *Gourmet's Guide to the Mind,* Belgium: Chromatika.
Brown, J.W. (2012) *Love and Other Emotions,* London: Karnac.
Brown, J.W. (2014) Mind and brain: a contribution from microgenetic theory, *Journal of Consciousness Studies,* 21, pp. 54–73.
Brown, J.W. (in press) Review of M. Weber and A. Weekes (eds.) *Process Approaches to Consciousness in Psychology, Neuroscience and Philosophy of Mind, Journal of Consciousness Studies.*
Brown, J.W. & Pachalska, M. (2003) The nature of the symptom and its relevance for neuropsychology, *Acta Neuropsychologica,* 1, pp. 1–11.
Bruner J., Goodnow, J. & Austin, G. (1956) *A Study of Thinking,* New York: Wiley.
Caird, J. (1880/1904ed) *An Introduction to the Philosophy of Religion,* Glasgow: James Maclehose & Sons.
Cajal, R. (1954) *Neuron Theory or Reticular Theory,* Madrid: Consejo Superior de Investigaciones Cientificas.
Cobb, J. (2008) *Whitehead Word Book,* Claremont, CA: P&F Press.
Collingwood, R. (1940) *An Essay on Metaphysics,* Oxford: Clarendon Press.
Coomaraswandy, A (1993ed) *Time and Eternity,* New Delhi: Munshiram Manoharlal.
Danto, A. (1981) *The Transfiguration of the Commonplace,* Cambridge, MA: Harvard University Press.
Dart, R. (1956) The relationships of brain size and brain pattern to human status, *S. Afr. J. Med. Sci.,* 2, pp. 23–45.
Davidson, D. (1980) *Essays on Actions and Events,* Oxford: Oxford University Press.
Dewan, E. (1976) Consciousness as an emergent causal agent in the context of control systems theory, in Globus, G., Maxwell, G. & Savodnik, I. (eds.) *Consciousness and the Brain,* New York: Plenum.
Dewey, J. (1925) *Experience and Nature,* Chicago, IL: Open Court.
Dickinson, G. (1937) *The Meaning of Good: A Dialogue,* Oxford: Clarendon Press.
Dombrowski, D. (2001) The replaceability argument, *Process Studies,* 30, pp. 22–35.
Eagleman, D. (2011) *Incognito: The Secret Lives of the Brain,* New York: Random House.
Ebbeson, S. (1984) Evolution and ontogeny of neural circuits, *Behavioral and Brain Sciences,* 7, pp. 321–366.
Eccles, J. (1970) *Facing Reality,* Berlin: Springer.
Eccles, J. (1992) Evolution of consciousness, *Proc. National Academy of Sciences,* 89, pp. 7320–7324.
Englefield, R. (1985) *The Mind at Work and Play,* Buffalo, NY: Prometheus.

Filk, T. & von Muller, A. (2009) Quantum physics and consciousness, *Mind and Matter*, 7, pp. 59–79.
Goodwin, B. (1982) Development and evolution, *Journal of Theoretical Biology*, 97, pp. 43–55.
Goodwin, B. (1984) Changing from an evolutionary to a generative paradigm in biology, in Pollard, J. (ed.) *Evolutionary Theory: Paths into the Future*, pp. 99120, New York: Wiley.
Gould, S. (1977) *Ontogeny and Phylogeny*, Cambridge, MA: Harvard University Press.
Gould, S. (1987) *Time's Arrow, Time's Cycle*, Cambridge, MA: Harvard University Press.
Griffin, D. (2001) *Animal Minds*, Chicago, IL: Chicago University Press.
Gunter, P. (1999) Bergson, mathematics and creativity, *Process Studies*, 28, pp. 268–288.
Guyau, J.-M. (1988ed) *Guyau and the Idea of Time*, Michon, J. (trans.), Amsterdam: North Holland.
Hartshorne, C. (1962) *The Logic of Perfection*, La Salle, IL: Open Court.
Hecaen, H. & Ropaert, R. (1959) Hallucinations auditives au cours de syndromes neurologiques, *Annales Medico-Psychologiques*, 117, pp. 257–306.
Heidegger, M. (1959) *An Introduction to Metaphysics*, Manheim, R. (trans.) New Haven, CT: Yale University Press.
Herrick, C. (1924) *Neurological Foundations of Animal Behavior*, New York: Henry Holt.
Humphrey, G. (1963) *Thinking*, New York: Wiley.
Hunt, H. (2001) Some perils of quantum consciousness, *Journal of Consciousness Studies*, 8, pp. 35–46.
Inge, W. (1924) *Personal Idealism and Mysticism*, London: Longmans, Green.
Isaacson, R. & Spear, N. (eds.) (1982) *The Expression of Knowledge*, New York: Plenum.
James, W. (1890) *Principles of Psychology*, Boston, MA: Holt.
James, W. (1907) *Pragmatism*, New York: Longmans, Green.
James, W. (1912) *Essays in Radical Empiricism*, New York: Longmans, Green and Co.
Kahneman, D. (2011) *Thought Fast and Slow*, New York: Ferrar, Straus and Giroux.
Kant, I. (1780–81) *Lectures on Ethics*, Ne wYork: Century.
Katz, M. (1983) Ontophyletics: Studying evolution beyond the genome, *Perspectives in Biology and Medicine*, 26, pp. 323–333.
Kelly, E. (2007) Psychophysiological influence, in Kelly, E., et al. (eds.) *Irreducible Mind*, pp. 117–239, Lanham, MD: Rowman and Littlefield.

Kim, J. (1993) The non-reductivist's troubles with mental causation, in Heil, J. & Mele, A. (eds.) *Mental Causation*, Oxford: Clarendon Press.

Kivy, P. (1993) *The Fine Art of Repetition: Essays in the Philosophy of Music*, Cambridge: Cambridge University Press.

Koestler, A. (1964) *The Act of Creation*, New York: Macmillan.

Köhler, W. (1938) *The Place of Value in a World of Facts*, New York: Liveright.

Laird, J. (1929) *The Idea of Value*, Cambridge: Cambridge University Press.

Lashley, K. (1950) In search of the engram, *Society of Experimental Biology Symposium Issue No. 4: Animal Mechanisms in Behavior*, Cambridge: Cambridge University Press.

Lashley, K. (1964ed) *Brain Mechanisms and Intelligence*, New York: Hafner.

Lenneberg, E. (1967) *Biological Foundations of Language*, New York: Wiley.

Levick, S. (1986) Unilateral auditory occlusions and auditory hallucinations, *British Journal of Psychiatry*, 148, pp. 747–748.

Lévy-Brühl, L. (1935/1983) *Primitive Mentality*, St. Lucia: University of Queensland Press.

Libet, B. (1985) Unconscious cerebral initiative and the role of conscious will in voluntary action, *Behavioral and Brain Sciences*, 8, pp. 529–566.

Libet, B. (1999) In Libet, B., Freeman, A. & Sutherland, K. (eds.) *The Volitional Brain: Towards a Neuroscience of Free Will*, Exeter: Imprint Academic.

Llewellyn, S. (2011) If waking and dreaming consciousness became de-differentiated, would schizophrenia result?, *Consciousness and Cognition*, 20, pp. 1059–1083.

Lucas, G. (1995) Whitehead and Wittgenstein, in Hintikka, J. & Puhl, K. (eds.) *Wittgenstein and the 20th Century Tradition of British Philosophy*, Vienna: Holder-Pichler-Tempsky Verlag.

Lucas, J. (1970) *The Freedom of the Will*, Oxford: Clarendon Press.

Luria, A. (1962) *Higher Cortical Functions in Man*, New York: Basic Books.

Luria, A. (1968) *The Mind of a Mnemonist*, New York: Basic Books.

Luria, A. (1976) *Basic Problems of Neurolinguistics*, The Hague: Mouton.

Lyons, W. (1986) *The Disappearance of Introspection*, Cambridge, MA: MIT Press.

MacLean, P. (1972) Cerebral evolution and emotional processes, *Annals of the New York Academy of Sciences*, 193, pp. 137–149.

MacLean, P. (1991) Neofrontocerebellar evolution in regard to computation and prediction: Some fractal aspects of microgenesis, in

Hanlon, R. (ed.) *Cognitive Microgenesis,* pp. 3-31, New York: Springer.
Martin J. (1972) Rhythmic (hierarchical) versus serial structure in speech and other behaviors, *Psychological Review,* 79, pp. 487-509.
McCullough, W. (1965) *Embodiments of Mind,* Cambridge, MA: MIT Press.
McDowell, J. (1994) *Mind and World,* Cambridge, MA: Harvard University Press.
McTaggart, J.M. (1934/1968) *Philosophical Studies,* New York: Books for Libraries Press.
McTaggart, J.M. (1901) *Studies in Hegelian Cosmology,* Cambridge: Cambridge University Press.
Mele, A. (1987) *Irrationality,* Oxford: Oxford University Press.
Merleau-Ponty, M. (1962ed) *Phenomenology of Perception,* London: Routledge.
Merleau-Ponty, M. (1964/1968ed) *The Visible and the Invisible,* Lingis, A. (trans.), Evanston, IL: Northwestern University Press.
Michon, J., Pouthas, V. & Jackson, J.L. (1988) *Guyau and the Idea of Time,* Amsterdam: Royal Netherlands Academy.
Pachalska, M., MacQueen, B.D. & Brown J.W. (2012) Microgenetic theory: Brain and mind in time, in Rieber, R.W. (ed.) *Encyclopedia of the History of Psychological Theories,* XXVI, pp. 675-708, Frankfurt: Springer.
Pachalska, M. & Weber, M. (eds.) (2008) *Neuropsychology and Philosophy of Mind: Essays in Honor of Jason W. Brown,* Frankfurt: Ontos Verlag.
Pachalska, M., MacQueen, B.D., Kaczmarek, B.L.J., Wilk-Franczuk, M. & Herman-Sucharska, I. (2011) A case of "Borrowed Identity Syndrome" after severe traumatic brain injury, *Medical Science Monitor,* 17 (1), pp. 18-28.
Pachalska, M. & Kropotov J.D. (eds.) (2014) *Psychology, Neuropsychology and Neurophysiology: Studies in Microgenetic Theory: In honor of Jason W. Brown,* Frankfurt/Lancaster, Ontos Verlag.
Pears, D. (ed.) (1963) *Freedom and the Will,* London: Macmillan.
Perry, R. (1926) *General Theory of Value,* New York Longmans, Green.
Peterson, E. (2011) The excluded philosophy of evo-devo? Revisiting Waddington's failed attempt to embed Whitehead's "organicism" in evolutionary biology, *Hist. Phil. Life Sci.,* 33, pp. 301-320.
Polanyi, M. (1958) *Personal Knowledge,* Chicago, IL: University of Chicago Press.
Polanyi, M. (1973) *Knowing and Being,* Chicago, IL: University of Chicago Press.
Pribram, K. (1991) *Brain and Perception,* Hillsdale, NJ: Erlbaum.

Pribram, K. & Luria, A.R. (1973) *Psychophysiology of the Frontal Lobes*, New York: Academic Press.

Rapaport, D. (1950) *Emotions and Memory*, New York: International Universities Press.

Richards, W. (1973) Time reproductions by H.M., *Acta Psychologica*, 37, pp. 279-282.

Romijn, H. (2002) Are virtual photons the elementary carriers of consciousness?, *Journal of Consciousness Studies*, 9, pp. 61-81.

Roth, P. (2008) *Indignation*, New York: Houghton Mifflin Harcourt.

Russell, B. (1914) *Our Knowledge of the External World*, London: George Allen & Unwin.

Russell, B. (1948) *Human Knowledge: Its Scope and Limits*, New York: Simon and Schuster.

Rusu, B. (2013) Feeling: from British Psychology to Whitehead's metaphysics, presentation for the conference on *Process Approaches to Consciousness*, Fontarèches, France, 4-6 April.

Santayana, G. (1923) *Skepticism and Animal Faith*, New York: Scribners Sons.

Schactel, E. (1947) Memory and childhood amnesia, *Psychiatry*, 10, pp. 1-26.

Schacter, D. (1985) Multiple forms of memory in humans and animals, in Weinberger, N., et al. (eds.) *Memory Systems of the Brain*, New York: Guilford.

Schilder, P. (1935) *The Image and Appearance of the Human Body*, London: Kegan.

Schilder, P. (1936) Psychopathology of time, *Journal of Nervous and Mental Disease*, 83, pp. 530-546.

Schilder, P. (1950) *Image and Appearance of the Human Body*, New York: IUP.

Schilder, P. (1951) On the development of thoughts, in Rapaport, D. (ed.) *Organization and Pathology of Thought*, pp. 497-519, New York: Columbia University Press.

Schilder, P. (1964) *Contributions to Developmental Neuropsychiatry*, New York: IUP.

Sherrington, C. (1951) *Man on his Nature*, Cambridge: Cambridge University Press.

Shields, G. (2009) Panexperientialism, quantum theory, and neuroplasticity, in Weber, M. & Weeks, A. (eds.) *Process Approaches to Consciousness in Psychology, Neuroscience, and Philosophy of Mind*, pp. 235-259, Albany, NY: SUNY Press.

Sprigge, T. (1993) *James and Bradley: American Truth and British Reality*, Chicago, IL: Open Court.

Sprigge, T. (2005) *The God of Metaphysics*, Oxford: Oxford University Press.
Stenner, P. (2008) A.N. Whitehead and subjectivity, *Subjectivity*, 22, pp. 90–109.
Stout, G.F. (1902) *Analytic Psychology*, London: Swan Sonnenschein.
Stout, G.F. (1921ed) *A Manual of Psychology*, London: University Tutorial Press.
Strawson, G. (2003) Real materialism, in Antony, L. & Hoernstein, N. (eds.) *Chomsky and his Critics*, pp. 49–88, Malsen, MA: Blackwell.
Strawson, G. (2006) Realistic monism: Why physicalism entails panpsychism, *Journal of Consciousness Studies*, 13, pp. 100–109.
Strawson, P. (1966) *The Bounds of Sense*, London: Methuen.
Streidter, G.F. & Northcutt, R.G. (1991) Biological hierarchies and the concept of homology, *Brain, Behavior and Evolution*, 38, pp. 177–189.
Stroud, J. (1956) The fine structure of psychological time, in Quastler, H. (ed.) *Information Theory in Psychology*, Wilmington, IL: Free Press.
Swedenborg, E. (1758/1940) *Heaven and its Wonders and Hell*, New York: Swedenborg Foundation.
Tucker, D. (2008) Self-organizing ontogenesis on the phyletic frame, in Pachalska, M. & Weber, M. (eds.) *Neuropsychology and Philosophy of Mind in Process*, pp. 371–400, Frankfurt: Ontos.
Vaihinger, H. (1924/1965) *The Philosophy of "As-if"*, Ogden, C.K. (trans.), London: Routledge & Kegan Paul.
Valéry, P. (1950ed) *Selected Writings*, Cowley, M. (trans.), Ne wYork: New Directions Press.
Vandervert, L. (1990) Symposium on a chaotic/fractal dynamic unification model for psychology, *Meeting of the American Psychological Association*, Boston, MA, 10–14 August.
Von Domarus, E. (1944) The specific laws of logic in schizophrenia, in Kasanin, J. (ed.) *Language and Thought in Schizophrenia*, Berkeley, CA: University of California Press.
Vygotsky, L. (1962) *Thought and Language*, Cambridge, MA: MIT Press.
Wallack, F. (1980) *The Epochal Nature of Process in Whitehead's Metaphysics*, Albany, NY: SUNY Press.
Ward, J. (1920) *Psychological Principles*, Cambridge: Cambridge University Press.
Weber, M. & Weekes, A. (2009) *Process Approaches to Consciousness in Psychology, Neuroscience, and Philosophy of Mind*, Albany, NY: SUNY Press.
Wegner, D. (2002) *The Illusion of Free Will*, Cambridge, MA: Bradford Press/MIT Press.

Weiszacker, V. von (1939-58) *Le cycle de la structure,* French translation of *Der Gestaltkreis,* Brudges: Desclée-de-Brouwer. (Original, 1939, Stuttgart: Thieme.)
Wertheimer, M. (1945ed) *Productive Thinking,* New York: Harper.
Whitehead, A.N. (1920) *The Concept of Nature,* Cambridge: Cambridge University Press.
Whitehead, A.N. (1929) *The Function of Reason,* Princeton, NJ: Princeton University Press.
Whitehead, A.N. (1933) *Adventures of Ideas,* Cambridge: Cambridge University Press.
Whitehead, A.N. (1954ed) *Dialogues, recorded by Lucian Price,* New York: Little, Brown.
Whitehead, A.N. (1978ed) *Process and Reality,* Corrected Ed., Griffin, D. & Sherburne, D. (eds.), New York: Free Press.
Whitrow, G. (1972) *What is Time?,* London: Thames and Hudson.
Whitrow, G. (1976) "Becoming" and the nature of time, in Čapek, M. (ed.) *The Concepts of Space and Time,* Amsterdam: Springer.
Williams, B. (1981) *Moral Luck,* Cambridge: Cambridge University Press.
Wittgenstein, L. (1953) *Philosophical Investigations,* Oxford: Basil Blackwell.
Wittgenstein, L. (1969) *On Certainty,* New York: Harper.
Yakovlev, P. (1948) Motility, behavior and the brain, *Journal of Nervous and Mental Disorders,* 107, pp. 313-335.
Yeats, W. (1932) After long silence, in *Words for Music Perhaps,* Dublin: Cuala.

Index of Names

Al-Azm, S. 19
Atmanspacher, H. 2, 26, 115, 117

Bachmann T. 4, 56
Bacon, J. 167
Bain, A. 125
Bartlett F. 106
Bender, M. 158
Bergson, H. 32, 37, 76, 89, 163, 169
Bernstein N. 18, 105
Bertalanffy, L. 115
Birch, C. 72
Boehme, J. 197
Bohm, D. 38
Borges, J. 109, 190
Bosanquet, B. 65
Bradley, F. 32, 34, 66, 67, 134, 187–94
Brentano, F. 92
Bricklin, J. ix, 20
Bruner, J. 106

Cajal, R. 16
Chomsky, N. 10
Cobb, J. 72, 171
Collingwood, R. 100, 140

Danto, A. 35, 57
Dart, R. 73
Davidson, D. 92
Dewan, E. 16, 30, 31
Dewey, J. 30, 31, 34, 69, 172
Dickinson, G. 142
Dombrowski, D. 72

Eagleman, D. 105
Ebbeson, S. 3, 5
Eccles, J. 9, 106, 118
Eckart, M. 199
Einstein, A. 47
Eliot, T. 189
Englefield, R. 102

Ewing, A. 1
Filk, T. 26, 115
Freud, S. 78, 101
Goethe, W. 5, 38, 65
Goldberg, G. 38
Goldschmidt, R. 73
Goodwin, B. 38
Gould, S. 3, 5, 73
Griffin, D. 8
Gunter, P. 74
Guyau, J.-M. 102

Hartmann, E. von 147
Hartshorne, C. 9
Hecaen, H. 141
Heidegger, M. 88, 106, 200
Herrick, C. 16
Hume, D. 57, 95, 102, 123
Humphrey, G. 94
Hunt, H. 72

Inge, W. 14
Isaacson, R. 120, 157

James, W. x, 9, 32, 35, 57, 73, 95, 113, 117, 143, 144, 164

Kahneman, D. 106
Kant, I. 19, 95, 154
Katz, M. 6
Kelly, E. 2
Kim, J. 114
Kivy, P. 43
Koestler, A. 143
Köhler, W. 5

Laird, J. 16
Lashley, K. 16, 31
Lenneberg, E. 73
Levick S. 96

Lévy-Brühl, L. 139
Libet, B. 20, 60, 105, 123, 124, 169
Linke, D. 153
Llewellyn, S. 55
Lucas, G. 138
Lucas, J. 135
Luria, A. 109, 120, 190
Lyons, W. 157

MacLean, P. 4, 118, 166
Martin, J. 18, 105
McCullough, W. 33, 145, 190
McTaggart, J.M. 63, 76. 117, 169
Mele, A. 116, 140
Merleau-Ponty, M. xii, 36
Michon, J. 102
Miller, G. 90
Moore, G. 141, 146

Nietzsche, F. 150, 182
Northcutt, R. 38

Pachalska, M. 3, 5, 56, 61
Pears, D. 160
Perry, R. 74
Peterson, E. 38
Polanyi, M. 111, 140
Popper, K. 182
Pribram, K. 3, 5, 15, 120

Rapaport, D. 68
Richards, W. 40
Romijn, H. 72
Roth, P. 191
Ropaert, R. 141
Russell, B. 11, 34, 98, 137
Rusu, B. 67

Schactel, E. 52
Schacter, D. 157
Schilder, P. 33, 40, 68, 94

Searle, J. 93
Shapley, H. 149
Sherrington, C. 16
Shields, G. 72
Spear, N. 120, 157
Sprigge, T. 32
Stout, G. vi, 12, 36, 155
Strawson, G. 24, 73
Strawson, R. xiii
Streidter, G. 38
Stroud, J. 20
Swedenborg, E. 148

Tennyson, A. 200
Tucker, D. 4, 106

Vaihinger, H. 154, 162
Valéry, P. 59
Vandervert, L. 4, 168
Von Domarus, E. 139
von Muller, A. 115
Vygotsky, L. 5. 59. 122

Waddington, 38
Wallack, F. 80, 119
Ward, J. 66
Weber, M. 4, 9, 56, 69
Weekes, A. 9
Wegner, D. 156
Weiszacker, V. von 17
Wertheimer, M. 16
Whitehead A.N. 4, 26, 36, 56, 66–9, 78,
    81, 83,112, 134–40, 169–71
Whitrow, G. 39, 169
Williams, B. 130, 155
Wittgenstein, L. viii, 10, 16, 53, 91,
    134–40, 144–51
Wordsworth, W. 19, 111

Yakovlev, P. 53, 105
Yeats, W. 87, 171

www.ingramcontent.com/pod-product-compliance
Lightning Source LLC
Chambersburg PA
CBHW051522230426
43668CB00012B/1700